How to Do
Ecology

How to Do
Ecology

A Concise Handbook

**Richard Karban
and
Mikaela Huntzinger**

Princeton University Press

Princeton and Oxford

Published by Princeton University Press, 41 William
Street, Princeton, New Jersey 08540
In the United Kingdom: Princeton University Press,
3 Market Place, Woodstock, Oxfordshire OX20 1SY

Library of Congress Cataloging-in-Publication Data
Karban, Richard.
How to do ecology : a concise handbook / Richard
Karban and Mikaela Huntzinger.
p. cm.
Includes bibliographical references and index.
ISBN-13: 978-0-691-12576-3 (hardcover : alk. paper)
ISBN-10: 0-691-12576-7 (hardcover : alk. paper)
ISBN-13: 978-0-691-12577-0 (softcover : alk. paper)
ISBN-10: 0-691-12577-5 (softcover : alk. paper)
1. Ecology—Research—Handbooks, manuals, etc.
2. Ecology—Experiments—Handbooks, manuals, etc.
I. Huntzinger, Mikaela, 1969– II. Title.
QH541.2.K374 2006
577.072—dc22   2006014269

British Library Cataloging-in-Publication Data is
available

This book has been composed in New Baskerville
Printed on acid-free paper. ∞

pup.princeton.edu

Printed in the United States of America

10 9 8 7 6 5 4 3 2 1

Contents

Illustrations

Boxes

The Aims of This Book

As students of ecology we take classes in ecological principles and ecological theory. We familiarize ourselves with the influential studies that have shaped our discipline. However, rarely do we explicitly try to figure out how to do ecology ourselves. What are the skills that are required to do a good job? How can we develop them? In a nutshell, this book is an attempt to provide a concise set of suggestions for how to do ecology well. This handbook is intended for students and practicing ecologists who are faced with developing an exciting research program.

In this handbook we will consider different ecological approaches and discuss their strengths, weaknesses, and utility. We will concentrate on experimental hypothesis testing, as this is the approach currently favored by most modern ecologists and as this is the approach that we are most familiar with. We will present some rules of thumb for how to set up experiments and how to analyze and interpret results. Finally, we will offer suggestions about working with other people and communicating what you find in scientific papers, talks, and proposals.

How to Do
Ecology

CHAPTER 1

Picking a Question

Perhaps the most critical step in doing field biology is picking a question. Tragically, as a young ecologist, it's the thing that you are expected to do first when you have the least experience. For example, it helps to get into grad school if you appear to be focused on a particular set of questions that matches a professor's interests. However, at this stage in most students' careers, many topics sound equally interesting, so this forced focus is difficult or even painful.

The question that you pick should reflect your goals as a biologist. If you are a new grad student, your short-term goal might be nothing more than to succeed in grad school. However, it's important to look farther down the road even as you're beginning. A common mid-term goal is getting your first job. For most jobs—those at research universities, small liberal arts colleges, federal agencies, nonprofit organizations—search committees all want to see a strong record of research and publication even if you will not be expected to conduct research or publish a lot on the job. Box 1 presents a justification for this bias. Search committees want to know that you are capable of advancing the field and

Box 1. *The importance of research for people who aspire to nonresearch careers*

Even if a career in research is not part of your long-term goals, it is still worth throwing yourself into the world of research while you work on your degree. The process of doing research will give you insights into ecology that are extremely difficult to get anywhere else.

- Doing experiments yourself helps you understand how individual biases, preconceptions, and points of view shape the ecological information that appears in textbooks.
- Over time, working on independent research helps you to incorporate the scientific method into your own thinking, which allows you to analyze reports and articles critically and to teach the information to others more effectively.
- Writing a thesis teaches even strong writers how to write more efficiently, concisely, and clearly.

These and other insights and skills are virtually impossible to gain solely through reading; instead, you are more likely to learn these things by truly immersing yourself in your research. And besides, it's fun.

communicating effectively. (They may also want to see other qualifications and experiences, such as teaching, etc.) Being a productive researcher demands a mid-term plan for your research. For example, this might include solving a particular problem in conservation, such as whether a single large or several small reserves are more beneficial for amphibian diversity in your region. More conceptual mid-term goals might involve making people rethink the interactions that are important determinants of the abundance or distribution of species.

Long-term goals are harder to formulate but are at least as important. (If you don't believe this, talk to some burnt-out researchers late in their careers. Some people never bothered to stop and figure out what they really valued and wanted to accomplish for themselves.) You should push yourself to pose a question that satisfies your goals and will be of broad interest. Some long-term goals that you might want to try out include attempting to influence how we think about or practice a subdiscipline of biology or how we manage a habitat or a crop. Having these goals in mind can provide a yardstick with which to evaluate your choice of project. In other words, figure out what you care the most about before picking a project. Do you most want to save sea turtles or to find a general ecological law? A project that is exciting to someone interested in trying to save a piece of the planet may not be satisfying to someone else who is trying to change how ecologists think about larval recruitment.

Your choice of goal should suit you and not necessarily your major advisor (who may consider a nonacademic career a waste of time) and not necessarily your parents (who may try to convince you that a conceptual thesis will leave you unemployable). That said, you should also recognize that if you answer a very specific question, your results may be considered important by a very small community. Academics are more likely to get enthused about a more general question. If your question is too general (theoretical), ask yourself if it reflects reality for at least one actual species. Having a model organism in mind will keep you more grounded in reality and increase the size of your audience. If your question is specific, ask whether you can generalize from your results. For example, you may choose to determine whether a specific disease causes symptoms in grapevines that are distinguishable from those caused by water stress. The answer to this question may be of considerable interest to grape growers but to no one else. Perhaps you can also ask the broader question of whether disease causes different plant responses than abiotic stress. The answer to this question will be interesting to a wider audience. Sometimes your funding may come from an applied source requiring that you answer a specific question about fisheries biology, restoration, and so on. It may not be possible to couch your question in more general terms. If so, you may be able to ask a complementary, parallel question that is more conceptual.

All projects have to be novel and original to some extent. You can't repeat work that has already been done and expect anyone to be excited about your results. We all like to hear new stories and new ideas and there is certainly a large premium placed on novelty. If you are asking the same question that has been answered in other systems, it behooves you to think about what you can do to set your study apart from the others. That said, if you are trying to start a project and you are stuck thinking up a novel idea, a useful way to begin may be to repeat an experiment or a study that captured your attention and imagination. Sometimes repeating a published study as a jumping off place will allow you to get started and move in an exciting new direction.

Policy makers are much less concerned with novelty than are academics, so what we just said about novelty may not apply as much if you are funded by an agency to answer a specific policy question. This means you will be asked to balance the expectation for novelty from your academic colleagues and the demands to answer the specific question for which your funding source is giving you money. Your first priority should be to generate relevant data for the agency; however, you should keep your eyes open for alternative answers and approaches. Asking additional questions in your study system that can lead to publishable research is also well worth considering.

Don't obsess about thinking up the perfect study before you are willing to begin (see box 2). One of the most unsuccessful personality traits in this business is perfectionism. Field studies are never going to be perfect. Don't get stuck thinking that you need to read more before you can do anything else. Reading broadly is great, but you will learn more by watching, tweaking,

Box 2. *Advice on picking questions for three types of ecologists*

There are three kinds of ecologists:
- The perfectionists who can't get started,
- The jackrabbits who have a lot of energy and want to get started before thinking through their goals, and
- Those who are just right, someplace in between.

Our advice differs depending upon where you fall on this continuum. If you are a perfectionist who can't get started because you haven't thought of the perfect question, we suggest you just get out there and do it. The experience and insight (not to mention publications) that you'll get by doing an imperfect study will help you improve in the future. If you are a jackrabbit and find yourself starting a million projects, our advice is to step back for a minute and ask which of these questions is most likely to advance the field and, even more importantly, to inspire enduring passion in you. And if you are a person who is just right, don't get a swelled head about it.

and thinking about your system. In addition, it is not realistic to expect yourself to sit at your desk and conjure up the perfect study that will revolutionize the field. Revolutionary questions don't get asked in isolation; they evolve. You start asking one question, hit a few brick walls, get exposed to some ideas or observations that you hadn't previously considered, and pretty soon you're asking very different questions that are better than your initial naïve ones. Most projects don't progress as we conceived them.

It is fine to start by asking a relatively "small" question. By small we mean specific to your study system and with relatively little replication. Small questions often generate more excitement than bigger ones because their more modest goals can be achieved with relatively few data, much sooner. Imagine that you want to study predation rates on goose eggs. These eggs are difficult to find and highly seasonal. You could conduct a small pilot experiment with three cartons of eggs from the grocery store. Your pilot study will not provide definitive answers about goose eggs but will likely provide useful insights about how to conduct an experiment for your project. If results from the pilot study turn out as expected, they can provide a foundation for a bigger project. If the results are unexpected, they can serve as a springboard for a novel working hypothesis. Almost all of our long-term projects had their beginnings as small pilot "dabbles."

Fieldwork is a hard business and many of the factors associated with failure or success are beyond your control. You should ask whether your ideas are feasible—are you likely to get an answer to the questions that you pose? Do you have the resources and knowledge to complete the project? To deal with the reality that field projects are hard to pull off, we suggest that you try several pilot studies simultaneously. If you know that you want to ask a particular question, try it out on several systems at the same time. You'll soon get a sense that the logistics in some systems are much more difficult than in others and the biological details make some systems more amenable to answering particular questions. It is a lucky coincidence that Gregor Mendel worked on peas, since they are particularly well suited to elucidate the particulate nature of inheritance. Other people attempted to ask similar questions but were less fortunate in the systems that they chose to investigate. Since most field projects don't work, have several possibilities that you try and follow the leads that seem the most promising. Don't get discouraged about the ones that don't work. Successful people never tell you about the many projects they didn't pull off. You should feel fortunate if two out of seven work well.

An essential ingredient of a good project is that you feel excited about it. The people who are the most successful over the long haul are those who work the hardest. No matter how disciplined you are, working hard is much easier if it doesn't feel like work but rather something that you are passionate about. As Confucius is supposed to have said, "Choose a job you love, and you will never have to work a day in your life." Pick a project that is intellectually stimulating *to you*. You are the one who has to be jazzed enough about it to want to do the boring grunt work that all field projects involve. You will feel much more inclined to stay out there in the pouring rain, through all the mind-numbing repetitions that are required to get a large enough sample size, if you have a burning interest in your question and your system.

There are two approaches to picking a project: starting with the question or starting with the system. The difference between these two is actually smaller than it sounds because you generally have to bounce between both concerns to end up with a good question. So regardless of which you start with, you need to make sure that you are satisfying a list of criteria related to both.

Many successful studies start with a question. You may be interested in a particular kind of interaction or pattern for its own sake or because of its potential consequences. For example, you may be excited by the hypothesis that more diverse ecological systems are intrinsically more stable. Alternatively you may be interested in this hypothesized relationship because if it is true, it could provide a sound rationale for conserving biodiversity, and if it is not generally true, ecologists should not attempt to use it as a basis for conservation policy. Since many studies have considered this question, you should think about what's at the basis or core of the hypothesis and whether previous studies have addressed these key elements. Are there novel aspects of this question that haven't been addressed yet? Even questions that have been addressed by many researchers may still have components that have yet to be asked.

If you start by asking a question, you will need to find a suitable system to answer it. The system has to be conveniently located and common enough for you to get enough replication. Ideally, it should be protected from vandalism by curious people and animals (or it should be possible for you to minimize these risks). It should be amenable to the manipulations that you would like to subject it to and apparent enough for your observations. You can get help finding systems by seeing what similar studies in the literature have used, by asking around, or by looking at what's available at field stations or other protected sites close to your home. The appropriate system will depend upon the specific questions that you want to ask. If your question requires you to know how your treatments affect fitness, you will want to find some annual rather than a charismatic but long-lived species. If your hypothesis relies upon a long history of coevolution, you should probably consider native systems rather than species that have been recently introduced. (Incidentally, there is a widespread chauvinism about working in pristine ecosystems. The implicit argument seems to be that the only places where we can still learn about nature are those that have not been altered by human intervention. We wonder if any such places really exist. Certainly less disturbed places are inspiring and fun, but they also represent a very small fraction of the earth's ecosystems. There are still plenty of big questions about how nature works that can be worked out in your own backyard regardless of where you live—we can attest to this, having lived in some uninspiring places.) One danger to guard against is trying to shoehorn a system to fit your pet hypothesis. If you start with a question, make sure you are willing to look around for the right system for that question and that you are willing to modify your question as necessary to go where the natural history of your chosen system takes you. You cannot make your organisms have a different natural history, so you must be willing to accept and work with what you encounter.

If you start with a system because of your interests, your funding, your major professor, whatever, you may find yourself in search of a question. Often one organism becomes a model for one kind of question, but it has not been explored for others. For example, the ecologies of lab darlings *Drosophila* and *Arabidopsis* are poorly known in the field. If everyone has used a system to ask one kind of question, there may be a lot of background natural history known about that system, but nobody has thought to ask the questions that you have. If you have a system but need a question, try reading broadly to get a sense of the kinds of questions that are exciting and interesting to you.

If you don't already have a system in mind but want to use this approach, try going to a natural area and spending a few days just looking at what's there. Generate a list of systems and questions in your notebook that you can mull over and prioritize later. Another useful approach is to start with a natural pattern that you observe. First quantify that pattern. You might observe that snails are at a particular density at your study site. Next ask whether there is natural variation in this measurement. Do some microhabitats have more snails than others? Is there natural variation that is associated with behavioral traits? For example, are the snails in some spots active but those in other places aestivating? Is there variation among individuals? Are the snails in some microenvironments bigger than others? Are bigger snails more active? And so on. Once you have quantified these patterns, ask: (1) what mechanisms could cause the patterns that you observe? And (2) what consequences could the patterns have on individuals and on other organisms? Even if a pattern you observe in your scouting has been described before, there are likely to be many great projects available. If it is an important and general pattern, it has probably been described by many people. However, it is less likely that the ecological mechanisms that cause the pattern have been evaluated. Understanding ecological mechanisms not only provides insight into how a process works, but also can tell us about its effects and where we would predict it to occur. Elucidating the mechanisms of a well-known pattern is likely to be a valuable contribution. Generate a list of potential mechanisms and then devise ways to collect evidence to test the strength of each of these. It is also less likely that its consequences have been described. Is the pattern important? Does it affect the fitness of the organisms that show it? Does it affect their population dynamics? Does it affect the behaviors of organisms that interact with it? Answering any one of these questions is plenty for a dissertation.

Don't assume that questions have been answered just because they seem obvious. For example, thousands of studies have documented predation by birds on phytophagous insects, but the effects of that predation on herbivory and plant fitness have only begun to be explored and are still very poorly known (Marquis and Whelan 1995). Although periodical cicadas are the most abundant herbivores of eastern deciduous forests of North America, their interactions with their host plants and the rest of the community are largely unexplored (Yang 2004).

Sometimes ecologists are constrained by funding sources or by labs that work on one set of organisms. If so, all of the obvious questions may appear to have already been addressed. Again, consider asking questions about the ecological consequences of what everyone else works on. For example, if you work in a lab where everyone works on the morphological changes in an herbivore that are induced by exposure to various predators, one more demonstration of an induced response may not be very novel. Perhaps you can ask what the fitness consequences of the different morphologies may be. Alternatively, try turning the question on its head and ask how predators and competitors respond to different morphologies of the herbivores.

Once you have selected a question and collected some preliminary data so that you know it is feasible to answer the question, next think about how to answer it as completely as possible. Here are some additional questions that could make your study more complete.

1. Consider alternative hypotheses to produce the patterns and results that you observe (see chapter 3).

2. Think about whether the phenomenon that you are studying applies generally. For instance, you may want to repeat your studies that gave interesting results at other field sites. You might also want to repeat them with other species.

3. Explore whether your phenomenon operates at realistic spatial and temporal scales. For instance, if you conducted a small-scale experiment, do your results apply at the larger scales where the organisms actually live (see chapter 3)?

4. If possible, work at levels both below (mechanisms) and above (consequences) the level of your pattern. For example, think about the ecological mechanisms that could generate the pattern that you observe. Again, understanding the mechanisms will help you to predict when and where to expect the phenomenon. Think also about the potential ecological consequences of your phenomenon. What other organisms or processes could it affect?

You may not be able to answer all of these questions, but the more complete your story is, the more useful and appreciated your work is likely to be. Each of these additional questions can take a lot of time and energy, so don't necessarily expect to address them all.

No matter how you first get started doing field biology, allow your organisms to redirect your questions. Many discoveries in science are unplanned. While you are answering one question, you are likely to see things that you haven't imagined. There is some chance that nobody else has seen them either. Rather than trying to force your organisms to answer your questions, allow them to suggest new ones to you. Read broadly so that you recognize that something is novel when you stumble upon it. Above all, be opportunistic!

CHAPTER 2

Posing Questions (or Picking an Approach)

Much of what you can learn about ecology depends on the questions that you ask. Your preconceptions and intuition determine the factors that you choose to examine, and these will constrain your results. Ecologists take several different approaches to science, and these approaches also constrain the kinds of answers that they get. Answers to the questions that you ask then form your view of how the natural world works. Deciding on an approach may sound like a bunch of philosophical nonsense to waste time, but it can have important consequences on everything that follows.

Different Ways to Do Ecology

Ecologists use several different approaches to understand phenomena, which we place in three categories: (1) observations of patterns, (2) manipulative experiments, and (3) model building. As is often the case in ecology, these categories are not mutually exclusive, and each has something to offer.

Observations of Patterns or Natural History

Observations of patterns in natural systems are essential, as they provide us with the players (factors and processes) that may be important. Observations allow us to generate hypotheses and to test models. Natural history used to be the mainstay in ecology, but it started to go out of style in the 1960s. Current training in ecology has become less and less based on a background in natural history (Futuyma 1998). Undergraduate education requires fewer hours of labs than it did in the past because labs are expensive and time consuming to teach. Traditional courses in the "ologies" (entomology, ornithology, herpetology, etc.) are becoming endangered. Graduate students are pressured to get started on a thesis project ASAP, often while they are still taking courses. This doesn't give them time to learn about real ecological systems before picking dissertation questions. The situation doesn't get any easier as students move on to become professors who are most "successful" by becoming research administrators. They write grants to fund other people to work with the organisms, allowing themselves more time to write papers, progress reports, and the next grants. As a result, the intuition for our experiments and models comes from the literature, the computer screen, or the intuition of our major professor. We spend a lot of time refining what everyone already believes is important. This has the danger of making our field conservative and unexciting.

It is clear to us that ecology as a discipline would be

improved if we were encouraged to learn more about nature by observing it first and manipulating and modeling it second. Observations are absolutely necessary to provide the insights that make for good experiments and models (Roush 1995). Experiments usually manipulate only one or a small number of factors because of logistical constraints. The factors that we as experimenters choose to manipulate determine the factors that we will conclude are important. For instance, if we test the hypothesis that competition affects community structure, we are more likely to learn something about competition and less likely to learn something about some other factor that we did not think to manipulate. Where do our choices of the factors that we consider come from, if not from personal experience with the organisms?

Good intuition is the first requirement for meaningful experiments. The best way to develop that intuition is by observing organisms in the field. Sadly, none of us "has the time" to spend observing nature. Guidance committees and tenure reviewers are not likely to recommend spending precious time in this way. However, observations are absolutely essential for you to generate working hypotheses that are novel, yet grounded in reality. Carve out some time to get to know your organisms. If you are too busy with classes and other responsibilities, then reserve two days before you start your experiments to observe your system with no manipulations—or preconceived notions. It often helps to do this with a lab mate, colleague, or captive family member. The opposite is also true; spending a whole day with no other people around and no distractions just looking at your organism can be very instructive as well. Even after you have set up your manipulations, continue to monitor the natural variation in your organisms. This will help you interpret your results and plan better experiments next year. For example, Mikaela's initial project plan for her first research project involved examining the role of fire on butterfly assemblages on forested hillsides. Poking around during her first season revealed that most butterflies were using riparian habitats, a microenvironment that fire ecologists had largely ignored. This led to a second experiment the following year that was far more informative than the original experiment she had planned (Huntzinger 2003).

Keeping a field notebook is one tool that may help you make and use observations. It is very difficult to remember the details that you observe. Jot them down in your notebook even if they don't seem particularly relevant to the question that you are addressing. Also jot down ideas that you have in the field about your study organism, other organisms that it may interact with, general ideas about how ecology works, even unrelated ideas that pop into your mind. It is amazing how valuable some of these observations can be at a later time, even if a lot of what you write doesn't seem worth much right then or doesn't ever prove useful.

An excellent way to begin a project is by observing and quantifying a pattern in nature, as we mentioned in the previous chapter. Common ecological patterns include changes in a trait of interest that varies over space or time. This could be anything from a trait of individuals (e.g., beak length) to a trait of ecosystems (e.g., primary production). First ask, how variable is the trait? Is there a real pattern to the variation over space or time? For example, are there large differences in primary production from one place to the next? What factors correlate with the variation that you observe? For example, do the differences in productivity follow a latitudinal gradient? What other factors also covary with this response variable (e.g., species diversity)? For example, what factors vary from the poles to the equator that could help to explain the observed pattern in diversity? It is often helpful to represent the pattern as a figure with one variable on the x-axis and the other on the y-axis. This representation allows you to get a sense of the pattern—how strong it is and whether the relationship between the two variables appears linear. At this point, an experiment will help to determine if the two variables are causally linked. If the relationship appears linear, then an experiment with two levels of the independent (predictor) variable may be appropriate. For example, if the relationship between the number of pollinators and seed set is linear, then an experiment with and without pollinators may be informative. If the relationship is nonlinear (let's say hump-shaped), then an experiment with only two levels (with and without pollinators) will not be as informative as an experiment with many levels of pollinators.

Observations are critical for meaningful experiments; in some cases they even replace experiments as the best way to gain ecological understanding. This is due in part to the unhappy fact that many processes are difficult to manipulate experimentally. Manipulative experiments must often be conducted on small plots and over short periods of time (Diamond 1986). However, important ecological processes often occur at scales that are large or have little replication. These processes may involve organisms that cannot be manipulated for ethical reasons. Other processes are simply hard to manipulate in any realistic way. For example, manipulations involving vertebrate predators are difficult to achieve with any realism. Their home ranges are often larger than the plots available to most investigators. Removing predators is often more feasible than adding them, although any density manipulations may be unethical. Observations are often possible in these and other situations when experiments are problematic. Observational experiments still require replication and controls to learn the most from them (see "Manipulative Experiments").

Although it's tempting to extrapolate our results from small-scale experiments to more interesting and realistic processes at larger scales, it is difficult to justify doing so. One partial solution to this dilemma is to observe processes that have occurred over larger spatial and temporal scales and ask whether these observations support our intuition based on prior knowledge, modeling, and experiments at small scales. Such observations are sometimes termed "natural experiments," since the investigator does not randomly assign and impose the treatments (Diamond 1986).

Why have observations lost currency compared to other methods of doing ecology? Observations can be applied to test hypotheses but they are poor at establishing causality. This is their main limitation. For example, we can observe that two species do not co-occur as frequently as we would expect. This may suggest that the two are competing. In the early 1970s everyone was "observing competition" of this sort because it made such good theoretical sense. However, the observed lack of co-occurrence could be caused by the two species independently having different habitat preferences that have nothing to do with current competition. Observations alone do not allow the causes of the pattern to be determined. We will have more to say about problems of both scale and causality later. In summary, observations provide natural history intuition at realistic scales so that important factors for manipulative experiments and modeling can be identified.

Manipulative Experiments

Manipulative experiments vary only one thing (or at most a few). The experimenter controls that variable. Since only one factor has been varied, if the experiment has been set up properly, any responses can be attributed to the manipulation. This is very powerful for establishing causality. Treatments should be assigned randomly to each replicate so that the treatments are interspersed (Hurlbert 1984). Statistical tests can then be used to evaluate the likelihood that the observed effect was caused by chance or by the manipulation. These issues will be considered in much more detail throughout this handbook, particularly later in this chapter and in chapters 3 and 4.

Model Building

Modeling is an attempt to generalize, to distill the cogent factors and processes that produce the behaviors, population dynamics, and community patterns that we observe. The strengths of the approach are that it applies generally to many systems and that it allows us to identify the workings of the important elements. Mathematical modeling forces us to be explicit about our assumptions and about the ways that we envision the factors (individuals, species, etc.) to be related. Since we often make these assumptions anyway, writing a model almost always focuses our thinking. We all use models to organize our observations, although these are usually verbal generalizations rather than mathematical equations. The act of writing an explicit model forces us to be more precise about the logical progression that produces generalizations. Models also allow us to explore the bounds of the hypothesis. In other words, under what conditions does the hypothesis break down?

Models can be general or specific; both kinds are usually constructed of mathematical statements. General models allow us to formulate the logical links between variables. Specific models involve measured parameters from actual organisms and allow us to make detailed predictions (e.g., how much harvesting can a population sustain?).

Successful models can let us develop new hypotheses about how nature works or about how to manage ecological systems. Take the model of apparent competition, where two species at the same trophic level appear to be competing, although one species is actually causing shared predators to become more abundant, which depresses the second species (figure 1). Theoretical models predicted that apparent competition should be common in nature (Holt 1977). Partly as a result of these modeling results, Holt and others have looked for this phenomenon in nature, and it is indeed widespread (reviewed by Holt and Lawton 1994). Models have also proven useful in designing conservation and management strategies. For example, a detailed demographic model of declining loggerhead turtles indicated that populations were less sensitive to changes in mortality of eggs and hatchlings and more sensitive to changes in mortality of older individuals than had been realized previously (Crouse et al. 1987). This result prompted changes in the efforts to protect turtle populations, presumably for the better.

In addition to helping us develop new hypotheses, models can tell us where to look. Darwin observed many finches with different morphologies and life histories on his visit to the Galapagos. However, he didn't record which morphologies were found on which islands. This failure wasn't because he was a second-rate naturalist but rather because he had not yet generated a model of differentiation and speciation. As a result, information about the spatial patterns of morphology would not have seemed relevant.

Models can make logical connections easier to see. Often the consequences or results are well known and very visible but the processes that caused those results are difficult to assess. Models can force us to consider alternative mechanisms when the currently favored explanation does not produce the "right result" in our modeling effort. For example, Maron and Harrison (1997) were faced with trying to explain why high densities of tussock moth caterpillars were tightly aggregated. They knew from a caging experiment that the caterpillars could survive outside of the aggregation, although in nature the caterpillars were restricted to the aggregation area. Spatial models suggested that very patchy distributions could arise within homogeneous habitat if predation was strong and the dispersal of the moth was limited. As a result of these model predictions, they looked for this counterintuitive explanation and found that it was indeed consistent with their observations and experimental results.

It is unfortunate that modeling and natural history often attract different people, with different skills, and with little appreciation for the other's approach.

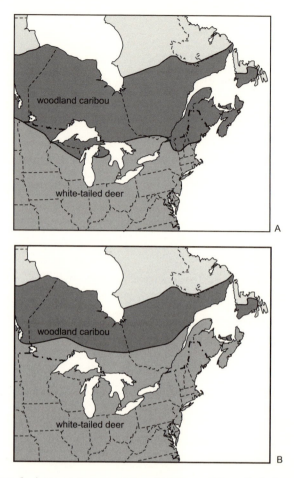

FIGURE 1. Apparent competition between white-tailed deer and caribou (Bergerud and Mercer 1989). (A) Caribou historically lived in New England, Atlantic Canada, and the northern Great Lakes states (redrawn from www.wildlandsleague.org). (B) Following European colonization, deer colonized these areas and replaced caribou (redrawn from Thomas and Gray 2002). Numerous efforts to reestablish caribou into areas where they could contact deer have failed. (C) The conventional hypothesis for caribou declines in locations with deer involved competition for food resources. More deer meant less food for caribou (shown as a nega-

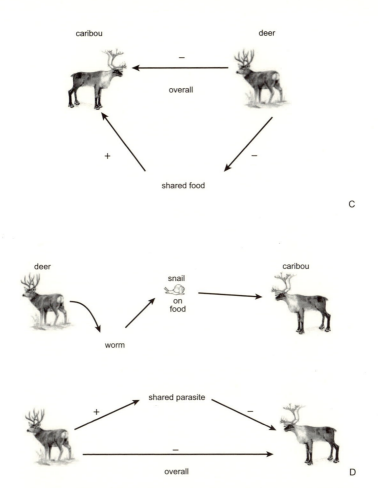

caribou

deer

−

overall

+

−

shared food

C

deer

snail

on
food

caribou

worm

shared parasite

+

−

−

overall

D

tive effect of deer on shared food). This explanation has not been supported by data, and one current hypothesis involves mortality to caribou caused by a shared parasite, a meningeal worm. (D) White-tailed deer are the usual host for the worm, and they are far more tolerant of infection than are moose, mule deer, and especially caribou. Caribou get the worms by ingesting snails and other gastropods that adhere to their food. The gastropods are an intermediate host for the worms. More deer cause more shared parasites, which reduce the number of caribou.

However, many of the most successful ecologists have been individuals who have been able to bridge these two approaches.

Why Ecologists Like Experiments So Much (or Why We Couldn't Call This Book "The Tao of Ecology")

The *Tao* is an ancient Chinese term that refers to the stream-like flow of nature. Like a stream, the Tao moves gently, seeking the path of least resistance and finding its way around, without disturbing or destroying. A Tao of Ecology might entail noninvasive and nondestructive observations of entire systems to understand who the players are and how they interact with one another. We find the Tao to be an appealing image in general and one that could be applied to ecology (see book cover). However, nothing could be farther from the approach that most ecologists currently employ. In this section, we explain why ecologists like to manipulate their systems so much.

In recent decades, ecology has come to rely on manipulative experiments. The investigator disturbs the system and observes what effect this disturbance has. This experimental approach has the advantage of providing more reliable information about cause and effect than do more passive methods of study. Understanding cause and effect is critical, powerful, and much more difficult than it sounds.

Consider the inferences that can be drawn based on observations versus those based on experiments. Observations allow us to make logical connections based on correlations. However, correlations provide limited insight into cause-and-effect relationships. One version of the old adage says that correlation does not imply causation. Bill Shipley (2000) points out that this is incorrect. Correlation almost always *implies* causation, but by itself, cannot *resolve which* of the two correlated variables might have *caused* the other. Let us give two examples, the first from one of our life experiences and the other from the ecological literature. The end of grad school was a time of reckoning for Rick. The only car he had ever owned, a Chevy Vega, was clearly falling apart, although he pretended not to notice. His girlfriend convinced him that since he had a job lined up on the other side of the country, and he would soon actually have a salary, he should abandon his grad-student life style and buy another car before heading west. Red has always been his favorite color, so naturally he was interested in a red car. However, his girlfriend had seen a figure on the front page of *USA Today* that red cars are involved in more accidents per mile than cars of other colors. Concerned about their safety, she argued for another color. Statistics don't lie, and red cars are more dangerous than other cars. Her working hypothesis had the cause and effect as "red causes danger":

$$Red \longrightarrow Danger$$

Rick was unsuccessful in convincing her that more dangerous (sexy?) people chose red cars in the first place and that getting a more boring color would do little to help them:

$$\text{Danger} \longrightarrow \text{Red}$$

In the end, Rick bought a grey car, but he drives a red one now (when a bicycle won't do). As this book goes to press he has luckily escaped being in any automobile accidents.

This example may seem silly, and scientists may seem unlikely to make this mistake (Rick's girlfriend was a social worker). We can assure you that we have seen it repeated many times by ecologists who infer causal links from correlations. For example, Tom White made the insightful observations that outbreaks of herbivorous psyllid insects were associated with physiological stress to their host plants and these outbreaks followed unusually wet winters plus unusually dry summers (White 1969):

$$\text{unusual weather} \underline{\hspace{1cm}} \text{physiological stress} \underline{\hspace{1cm}} \text{psyllid outbreaks}$$

He argued that plant physiological stress increased the availability of limiting nitrogen to the psyllids he studied, and to many other herbivores (White 1984). So essentially he hypothesized a causal connection between these correlated factors:

$$\text{weather} \longrightarrow \text{stress} \longrightarrow \text{increased N} \longrightarrow \text{herbivore outbreaks}$$

However, the actual causal links could be different. For instance:

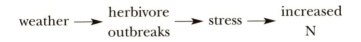

Or perhaps weather influences some other factor that then causes herbivore outbreaks, without involving the host plant:

weather ⇀ plant stress
         ↘ reduced predation ⇀ herbivore outbreaks

Without manipulative experiments, it is difficult to establish which of these causal hypotheses are valid and important. However, if microenvironmental conditions, physiological stress, available nitrogen, herbivore numbers, and predator numbers can all be manipulated, it will be relatively easy to determine which of these factors cause changes in which others. In the end, White's intuition got him fairly close to the truth. A recent review of experimental studies suggests that herbivores, especially the sap-feeders that White studied, are negatively affected by continuous drought stress, but intermittent bouts of plant stress and recovery promote herbivore populations (Huberty and Denno 2004).

Correlations do not specify causation. If plant stress is correlated with outbreaks of herbivores, we cannot know whether plant stress caused the outbreaks, the outbreaks caused the plant stress, or something else caused both of these results. Replicated manipulative experiments have the potential to provide more definitive evidence about causality, but unfortunately many ecological problems are not amenable to experimentation. New techniques are being developed that can provide inference about causal relationships from observational data (Shipley 2000). These techniques involve directed graphs (the diagrams with arrows shown above). Once we have specified a causal or directed graph, we can predict which pairs of variables will be correlated and which pairs will be independent of one another. This technique allows us to estimate the probability that the correlations or independencies have been generated by various causal models. Using this technique, we can discard causal models that don't fit our observations. Proposing a causal model that does not contradict the data does not mean that it is correct, since other models may also provide consistent results. But this method has the potential to provide a powerful way to infer causality from nonexperimental data. Bill Shipley provides a computer program that allows you to generate statistical tests of various causal models (Shipley 2000). These methods are not difficult to use, although they are much less well known than inferential statistics such as analysis of variance (see chapter 4). Shipley's approach is most useful when one model matches the observed patterns more accurately than alternative models; this is often not the case in ecology. Perhaps the biggest problem with these new techniques is that they depend upon a number of assumptions about causal structure. For example, it is assumed that there is no feedback between factors but rather, causality is asymmetrically directed (A causes B and B does not feed back to affect A). However, feedback of this sort is common in ecological systems and makes inference of causality much more difficult and murky. In summary, using directed graphs to infer causality without manipulative experiments is an exciting new technique, but the jury is still out on how often and severely the assumptions will limit the applicability of these new methods.

Models of all sorts have been used to test ecological hypotheses. If one or several models have been constructed that include particular ecological mechanisms, it is possible to ask whether the model fits actual data. If the fit is good, it is often argued that the ecological mechanisms included in the model are probably operating. However, this reasoning uses correlation to support the implicit hypothesis that the mechanisms that provide a good fit or correlation are the ones that caused the patterns in the data. Such arguments rarely acknowledge that other mechanisms could also produce good fits to the data and may be the actual causal factors in nature.

Ecologists love manipulative experiments because they are very powerful tools to establish cause and effect. Regardless of your philosophical persuasion about this issue, the truth remains that it is much easier to get

experimental work published and to get other ecologists to pay attention to it than if you are doing studies
relying on observations and correlations. However, experiments are only as good as the intuition that stimulated the experimenter to manipulate the few factors
that he or she has chosen. Experiments are limited by
this initial intuition and by problems of scale and realism. Natural history observations can provide the intuition to design meaningful experiments and provide information over larger areas and longer time frames than
an experimenter can handle with manipulations. Models can provide generality, suggest results when experiments are impossible, project into the future, and stimulate testable predictions.

Whenever possible, you should integrate several of
these approaches to pose and answer ecological questions. One approach can make up for the weaknesses
of another. The best of modern ecology combines observations, models, and manipulative experiments to
arrive at more complete explanations than any single
approach could provide.

CHAPTER 3

Using Experiments to Test Hypotheses

Manipulative experiments are the rage in ecology because they can help to establish causality. This comes about because the experimenter manipulates the treatments and observes the effects that the manipulations produce. In this and following chapters, we will consider experimental and statistical techniques that are used to evaluate cause-and-effect relationships in ecology.

Experimental Requisites

Unambiguous interpretation of causality is dependent on several requirements: (1) appropriate controls, (2) meaningful treatments, (3) replication of independent units, and (4) randomization and interspersion of treatments (Hurlbert 1984).

Controls

Controls are treatments against which manipulations are compared. Biological systems change over time. As a result, we cannot simply compare our experimental units before and after we apply treatments. Any differences that we observe before and after application of treatments could be caused by the treatments but also could be caused by other changes that occurred during this time period. Since both the controls and the manipulated treatments experience whatever changes occur over time, the effects caused by treatments can be separated from those caused by other changes. For example, we could experimentally add french fries, environmental estrogens, or something else to the diets of male deer during fall and winter. We would observe that they shed their antlers in March. Without controls we might conclude that the added experimental french fries had caused the antlers to be shed. However, controls without added fries would help us recognize that other seasonal changes were responsible for the antler shedding that we had observed. In this case, our control treatment is the population of male deer with no added food, essentially "no treatment."

The logical control in some experiments is not necessarily "no treatment." For example, if we wanted to evaluate effects of elevated CO_2 on plant growth, we might want to compare plants grown in an environment with CO_2 levels that are projected for 2050 with controls grown under current ambient conditions (no treatment). However, a more meaningful control might involve growing plants under conditions with 25% less CO_2 than current ambient levels, since this is the estimated level before the Industrial Revolution.

Meaningful Treatments

When we impose experimental treatments, we often change things other than the factors we are hoping to manipulate. For example, if we want to evaluate the effects of herbivores on plant traits, we might set up the following replicated and randomly assigned experimental treatments: (1) plants caged with herbivores and (2) control plants that lack cages and herbivores. This experiment seems straightforward and easy to interpret. However, any differences that we observe between these two treatments could be caused by the presence or absence of herbivores or could be caused by a variety of other confounding sources. The cages could produce effects that we erroneously assume to be caused by our treatments (presence or absence of herbivores): the cages may alter the microenvironment experienced by the plant; they may eliminate beneficial organisms (pollinators) from the plant; they may eliminate harmful organisms (plant pathogens; other herbivores) from the plant; the cages may interfere with the normal behavior of the herbivores so that their effect is greater or less than it would be without a cage; or they may interfere with the normal behavior of the plant so that its usual developmental or reproductive schedules are disrupted. The list goes on seemingly forever. The message is that establishing cause and effect by a simple manipulative experiment can be surprisingly difficult.

There are several ways to address the complications caused by manipulations that are intended to change only a single factor but actually change many additional and often unrecognized factors as well. One approach is to design a cage treatment that causes as few as possible of the unwanted, secondary effects described above. When this approach is not possible, these other effects can be measured to evaluate their likelihood as confounding artifacts. It may also be possible to include additional treatments that test for these potential artifacts. For example, cages can be constructed of different mesh sizes that allow some of the smaller organisms to come and go freely but block larger ones. Alternatively, cages can sometimes be left open at the top or bottom so that some of the microenvironmental side effects caused by the cage are included in the controls. It is often a good idea to attempt to impose the treatment in several different ways. Each of these different impositions may have its own side effects. So for instance, another way to exclude small herbivores is to treat plants with selective pesticides. Such a treatment is likely to cause its own artifacts. However, the artifacts associated with pesticides are probably different than those caused by caging. If you find that herbivores have a consistent effect on plants regardless of how they are experimentally manipulated, you can feel more confident that your conclusions are real and robust. Many of these potential problems of erroneously assuming that your treatments have caused your effects can be avoided by thinking about potential side effects of your treatments and attempting to include them in controls.

Real care should be taken in deciding on appropriate treatments and controls. Biological processes may or may not be easily mimicked by manipulative treatments. Consider the example of an experimental treatment that attempts to mimic the effects of fire on plants. Some ecologists mimic landscape-scale fires with small-scale experiments by conducting fires in 1 m² fireproof arenas placed around vegetation. These experimental fires only get a fraction as hot as real fires, they combust a small proportion of above-ground biomass, and so on. Similarly, the easiest and most straightforward way to mimic herbivores that chew foliage is to cut leaves with scissors. For some plants, clipping with scissors adequately captures the effects of herbivores. However, for other plants, how the leaf area is removed—in one big bite or many small ones, whether the veins are severed, etc—can greatly influence the effect of the removal (Baldwin 1988). In addition, components of saliva from particular herbivores have been found to have profound effects on some host plants (Felton and Eichenseer 1999). Carefully designed treatments involving actual organisms and actual processes, when feasible, are preferable to more artificial treatments.

When designing experimental treatments, it is important to attempt to span the natural range of variation. For instance, if we want to evaluate the effects of herbivores on plants, repeatedly defoliating the plants may produce dramatic effects. However, this result will tell us little about the effects of real herbivores on plants if repeated defoliation does not occur in nature. Conversely, picking only a single, modest level of damage may underestimate the effects of real herbivory. The best bet here might be to use a range of treatment levels that spans the range of damage that is naturally encountered.

Common sense and preliminary observations are usually better yardsticks for meaningful treatments than any that can be found in the literature. Nonetheless, you can get some useful ideas about methods in the following references: Moore and Chapman (1986) for sampling plants, Kearns and Inouye (1993) for methods used in studying pollination and plant fitness, Southwood and Henderson (2000) for sampling insects, and Wilson et al. (1996) for methods used to sample mammals.

Replication

It is important to replicate independent experimental units of each treatment and control so that you can separate the effects of the treatment from background noise. Imagine for a moment that there is only one replicate (independent sample) of each treatment. It will be impossible to determine if any differences between the treatments (manipulations) are really due to differences caused by the experimental manipulations or by other differences between the particular individuals sampled for each treatment. With only one independent replicate, no amount of subsampling or precision will help establish causality because the factors that affect one subsample may also affect others. However, if many independent replicates show a difference between treatments, we can be more confident that this difference was caused by the treatments. For example, Dave Reznick and John Endler (1982) observed that guppies from a Trinidadian stream that had high risk of predation from larger fish had different life histories than fish from a stream with low predation risk. Fish with high risk of predation became adults more quickly and devoted more resources to reproduction than those with low risk of predation. Collecting many guppies (subsamples) from one site of each treatment did little to improve the inference that predation was the cause of the life history differences. Instead Reznick found many different streams in Trinidad and characterized the life history traits of the guppies as well as their risk of predation. This gave him more confidence that the relationship between predation and guppy life histories was a real one. In addition, he performed a manipulative experiment, moving guppies from streams with high predation to streams with low predation (Reznick et al. 1990). Descendents of the transplanted guppies had traits that matched those that were found in low predation streams. This experiment was repeated in two different river systems.

Getting independent replicates is not always easy. In practice, try to separate independent replicates by enough space so that conditions at one replicate do not influence conditions at another replicate. This distance will be determined by the organisms in question—generally larger and more mobile organisms will need greater spacing than smaller or more sedentary ones. The point to remember is that in general, the more independent replicates you have, the greater is your power to detect treatment effects.

Often replication comes at the expense of precision of each estimate. This is fine. It is almost always better to take many samples, each with little precision, than to spend your limited time making sure that any given sample is accurate. The central limit theorem of probability can help you out here. If you take a large number of unbiased estimates, each one very imprecise, you will quickly arrive at a mean value that is close to the actual value. This is an impressive party trick. Get your friends to estimate the size of some object, say a window. Individual estimates are likely to be far from the actual value (man, do some people have bad judgment). However, the mean from a group of about 30 partyers will be astonishingly close to the actual value. The message from this exercise is clear. Always go for as large a sample size as you can get, even if each of your samples is sloppy and noisy. A large, unbiased set of samples will average over the noise and bale you out. This advice is supported by analyses of real ecological data: Zschokke and Ludin (2001) found that imprecise measurement had surprisingly little effect on ecological results, suggesting that limited time and resources are much better invested in more replicates than in more precision of measurements.

A large sample size can rescue imprecise measurements, but it cannot rescue biased measurements. For example, imagine asking a group of 30 partyers for their estimates of the age of the earth. The mean value will be quite different if the partyers are geologists or religious fundamentalists. Increasing the sample size will not alter this bias.

Students who are just starting to do research often want to know how large a sample size they will need. There is no easy answer to this question; it depends on the size of the difference that you are interested in detecting and how much noise there is. Yes, but let's get real. That knowledge doesn't help you determine how large a sample size to shoot for. As a general rule of thumb with no other information, we always try to get 30 independent replicates of each treatment. If 30 is impossible, 15 will do. Below 15, we start to get anxious.

Some experiments do not lend themselves to high replication. For example, conservation questions at the landscape scale can be replicated only a few times. We have conducted experiments in large plots that excluded different mammalian herbivores (Young et al. 1998, Augustine and McNaughton 2004). In these cases, each treatment was replicated only three times. Here subsampling allowed us to get a more precise estimate of the response in each plot so that statistically significant differences were revealed (Huntzinger et al. 2004, Huntzinger and Augustine, n.d.). It helped that the effects we were looking for were relatively huge; no amount of subsampling would have made these low levels of replication work for small effect sizes. Sometimes it is impossible to replicate your treatments at all. In these cases, statistical tests are probably inappropriate (although this suggestion is contentious—see Oksanen 2001). Results without statistical tests are difficult to publish by themselves but may accompany smaller-scale, replicated studies to provide biological realism to those statistically significant results.

On the subject of sample size, statisticians always recommend that you collect preliminary data to determine the appropriate sample size. They argue that doing this will save you time in the long run. Although this is sound advice, we have never heeded it. Will we ever conduct experiments this way? Maybe, but probably not before we retire and learn the virtue of patience.

Having a large number of replicates increases your power to detect differences caused by your treatments. However, high replication comes at the expense of the size of each sample. In other words, if you want to have many replicates, each of those replicates is going to be small; if you have fewer replicates, each one can occupy a larger area. This can be a serious issue because some processes only operate at particular spatial scales. For example, you are likely to get a different result placing a predator in a large arena where it can set up a normal territory and its prey can replenish themselves, than in a small arena where it will behave abnormally and quickly eliminate its prey.

Having replicates not only gives you statistical power but also makes common sense. With only one (or a few) replicates of each treatment, you have no idea if the differences that you observe were caused by your treatments or by some other factor for which the treatments happen to differ. Independent replicates that are not biased with respect to the treatments allow you to avoid spurious interpretations. However, insufficient replication is just one possible source of an incorrect interpretation. Experiments that are conducted using spatial and temporal scales that are too small can also lead to incorrect inference. Since replication almost always comes at the cost of scale, some ecologists argue that our field has leaned much too far in the direction of replication and that scale should take priority (Oksanen 2001). As previously mentioned, we often try to deal with this problem by working at two spatial scales—conducting a highly replicated manipulative experiment with small units, and a poorly replicated experiment or observation with large units. If the answers are similar at both scales, the conclusion is much stronger. This is useful when dealing with different audiences. Resource managers, growers, and agriculturists are not very hung up on statistical tests, but they won't listen to results conducted on small plots. On the other hand, most academics will not listen to results (or publish them) if they are not statistically significant, and this requires replication. The solution? You can't please these two audiences simultaneously. Conduct two different experiments, one with high replication and the other with a large spatial scale. It probably doesn't matter which scale you start with; there are unique advantages to both.

One way to visualize the scope of your experiments and observations is to plot them on a graph with spatial scale as one axis and temporal scale as the other axis (figure 2). In this way you can clearly see the spatial and temporal range that your manipulative experiments, natural experiments, and models cover. For example, Schneider et al. (1997) wanted to understand the multiannual population dynamics of a bivalve mollusc at the scale of an entire harbor (368 km^2). Their experimental units were 13 cm cores taken over a 30-second period. These units were repeated over an area of about half a km^2 during a 28-month period, which allowed them to greatly expand the scope of their experiment. Nonetheless, making inferences about dynamics that occur in the harbor on the scale of decades required considerable extrapolation. They combined this experiment with a model of the entire harbor. They used information from the small-scale experiment to suggest model parameters, and information from the model to help interpret their experimental results and suggest further experiments.

Since manipulative experiments in ecology today are almost always conducted at spatial and temporal scales

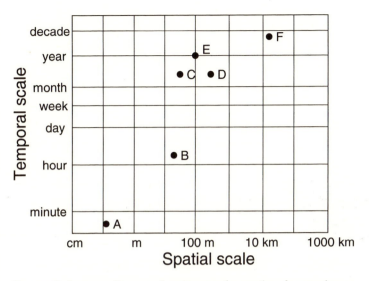

FIGURE 2. A scope diagram showing an observational, experimental, and modeling approach used to characterize changes that occur to harbors over decades (Schneider et al. 1997). The diagram makes explicit the spatial and temporal extent of each observation or experiment as well as the scale over which you would like to extrapolate. Each observational unit is a core with a 13 cm diameter taken over a 30-second period (A). A survey included 36 cores (B), at six sites (C), repeated during 12 monitoring visits (D). An experiment was conducted with a 13 cm core as the experimental unit (A). Ten cores were taken at each plot and there were 2 experimental sites each visited 9 times (E). Schneider et al. also constructed a model at the scale of the entire harbor (F) and attempted to extrapolate downward to compare their model results with those from observations (surveys) and experiments.

smaller than our ideal, it is worth considering what effect this has on our worldview. Small-scale experiments have led us as a group to believe in local determinism, that is, that the processes we can manipulate on a small scale mold the patterns that occur on larger scales (Ricklefs and Schluter 1993). However, this view is likely to be simplistic when we look at real communities. For example, local processes such as competition and predation tend to reduce species diversity, while larger-scale regional processes tend to increase diversity through movement and speciation. Our view of local determinism can be tested by placing a barrier around a local area and observing whether species persist. In most cases, they don't. Even the largest parks such as Yellowstone and the Serengeti are too small to maintain a full complement of species over the long term (Newmark 1995, 1996). An appreciation for larger-scale processes such as the interactions with other organisms in the species' range and interactions that occurred in the past can help our thinking, although these processes are difficult to study experimentally (Ricklefs and Schluter 1993, Thompson 1999).

Independent Replicates, Randomization, and Interspersion of Treatments

Replication only serves a useful purpose in experimental design if the replicates are spaced correctly. For instance, if all of your high-nitrogen replicates also happen to be in a swampy area and your controls are in a drier upland area, then you might conclude that nitrogen caused effects that were actually caused by the swamp. Therefore, experimental replicates must be independent of one another (Hurlbert 1984). Independent replicates make it likely that the noise associated with each treatment is unbiased and that the treatments are on average similar in all ways except for the treatment effect. Consider a common design for experiments that examine the effects of biological control agents (insect predators) on greenhouse pests. The greenhouse is physically divided in half with a screen barrier down the middle (figure 3A). Predators of the pest are released into one half of the greenhouse, containing, let's say, nine plants. No predators are released into the control half of the greenhouse, also containing nine plants. It is incorrect to assume that each of the two treatments is replicated nine times. If one side of the greenhouse is different from the other, then all of the plants of each treatment will experience those differences. In essence, there is only one independent replicate of each treatment.

The way to get independent replicates that are unbiased is to intersperse the replicates of the various treatments. In other words, placement of the two treatments must be all mixed up (figure 3B). If one side of the greenhouse is sunnier or windier than the other, these differences will not be confounded with treatment differences. Any observed differences associated with the treatments will probably be caused by the treatments. Of course, setting up this design is going to require many more screen barriers than the design that divides the greenhouse in half. In summary, by assigning treatments randomly, we can differentiate between

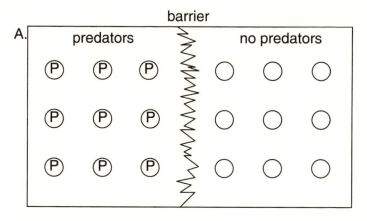

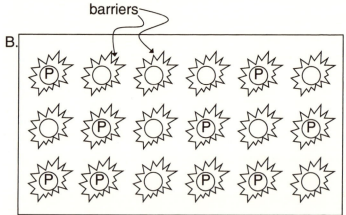

FIGURE 3. Experimental designs to evaluate the effects of predators on greenhouse pests. (A) A pseudo-replicated design with one
barrier separating the two treatments and only one independent
replicate of each treatment. Treatments are not interspersed. (B)
A proper design with 18 barriers and 9 independent replicates of
each treatment.

the possibility that our treatments caused the results that we observe versus the possibility that some other factor caused the results. If all of the replicates of the various treatments are randomly interspersed, then other uncontrolled factors are unlikely to cause the observed pattern. However, if all of the replicates of a treatment are grouped so that they share other factors, then we don't know if these uncontrolled factors or our treatment caused the pattern.

The best way to assign treatments is by using a random number generator. This is often most conveniently done at home before going out to the field. A deck of cards or a page ripped out of a telephone directory can provide random numbers (use the last four digits; the first three are not random). For two treatments, use red/black (cards) or even/odd (phone numbers), for three treatments use three suits or 1, 2, 3 numerals, making sure each treatment is assigned the same number of replicates. Random assignment does not mean assigning every other individual to each treatment, nor does it mean going along and haphazardly assigning treatments. Both of these methods are acceptable, but they should be identified as alternating or haphazard assignment of treatments, and both will reduce the power of your statistical inference. Randomization is an effective way to achieve interspersion of treatments. If your random assignment of treatments fails to provide treatments that are well interspersed and matched for other factors, the experimental units should probably be reassigned (Hurlbert 1984). In other words, if by chance many of one treatment wind up being on one side of the plot (which turns out to be drier) and many of the second treatment wind up being on the other side of the plot (which turns out the be wetter), moisture and treatment will be confounded. It is worth assigning treatments randomly a second time even if you don't suspect that the two sides differ in moisture or in some other unmeasured variable, since there are so many confounding spatial factors that you might not anticipate.

If you know that there is an environmental gradient in your study plot before you assign treatments, it is a good idea to *block* your experiment and assign treatments within blocks (Potvin 1993). For example, if you know that one side of your plot is on a slope and the other side is on level ground, divide (block) the field into two subplots (slope and level) and then randomly assign an equal number of replicates of each treatment to each of these two blocks. Blocking can reduce the noise caused by the environment and give you more power to detect an effect of your treatments than in a completely randomized design. However, there is a cost of blocking. Blocks decrease the error degrees of freedom in your analysis; the more blocks, the larger this effect. If blocking does not accurately match the environmental heterogeneity that is important to the outcome of your experiment, then blocking decreases power relative to a completely randomized design.

Many ecological studies are designed with inadequate replication and interspersion. Others have different flaws in their design, analysis, and interpretation. If you identify such a problem, it is a natural temptation to disregard or trash the results. We suggest that this temptation should be held in check. A study that has a design flaw may be subject to alternate explanations; on the other hand, the conclusions may be correct, and the study almost certainly has some biological intuition to offer. This is also true of your own studies. Too often people don't analyze data that they have collected because they feel their design wasn't perfect. This may be true, but the work almost certainly has something to teach you (and others). Let's keep things in perspective. Even the best experimental study is subject to alternative explanations because only a limited number of factors were considered. In chapter 2, we discussed Tom White's interpretation of the causal relationships between weather, plant stress, and outbreaks of psyllids. Since he only had a correlation, he couldn't establish cause and effect. Nevertheless, it would be a mistake to disregard his insights. Based on the information he had, he couldn't be sure of causality, but experiments done in the interim suggest that he was certainly on the right track (Huberty and Denno 2004). Without long-term data and biologists like Tom White who have an intuitive grasp of their organisms, we won't do the proper experiments, with or without statistical "rigor."

Lab, Greenhouse, or Field?

Ecologists are often tempted to work in environments that minimize unwanted variance or "noise." The reasoning is that controlled environments enable us to vary single factors in order to isolate their effects (Potvin 1993). Ecologists do this to varying degrees and in different ways. We choose field sites that are well matched and as similar to one another as possible. We move into the greenhouse, where abiotic conditions can be controlled and made similar for all replicates. Sometimes we conduct experiments in small growth chambers, aquaria, and lab "microcosms," which provide even more environmental control. The real world (field) can be so complicated that it is difficult or impossible to perceive patterns because of all the noise. A simplified, controlled environment can reduce this noise and allow us to see the signal, or test predictions, or get at mechanisms that would not be possible in the field. In addition, working in these controlled environments is often more convenient than working in the field. Controlled environments, such as greenhouses or growth chambers, are often close to where we have other obligations (e.g., our classes or families) and allow us to conduct experiments during times when natural systems may be inactive. They may also be close to where we have equipment that is essential to impose treatments or measure responses.

This control and convenience comes at a very large, and often unrecognized, cost. First, controlled environments are generally far more variable than we imagine (Potvin 1993). In our experience, plants and insects grow very differently on one side of a greenhouse bench than on the other side. This variance in the greenhouse is often larger than we have encountered in the field.

Second, working under controlled conditions is unrealistic. For example, plants in the greenhouse routinely experience outbreaks of pests that remain at quite low levels under field conditions. In addition, anything that you learn in the field is potentially interesting and important because it happened in the real world. It can lead to new research directions. For example, at the end of his thesis research, Rick set out to learn about mortality factors that could affect populations of cicadas in the field and noticed induced plant responses that killed cicada eggs (Karban 1983). This unexpected turn of events stimulated the questions about induced resistance that he asked for the next twenty-plus years. While asking questions in the field about induced resistance in wild tobacco plants, an unexpected frost damaged many of these plants. This seemed like a disaster at first, but Rick learned that induced plants are more susceptible to frost, and this risk may represent an unappreciated cost of induction (Karban and Maron 2001). In contrast, if the temperature controls don't work correctly in Rick's growth chamber, he won't learn anything useful about plant responses to real variation in temperature.

As a researcher, it is a good idea to be an opportunist and to go where your system takes you. This is much more difficult when working in the lab. For example, after comparing many repetitions of a lab experiment, it appeared that the strength of induced resistance varied from one experiment to the next. Some of this variation was due to using different pot sizes (Karban 1987). Plants in pots dry down at different rates, and pot-bound plants are less inducible. Who cares? This result provides very little inference about how organisms work in the real world.

Often ecological phenomena do not transfer from the field to the lab. For example, Henry Horn studies the mosses and lichens that grow on boulders near Princeton (personal communication). These small organisms survive droughts because the surface temperature of the large boulders lags behind the air on a daily cycle, distilling water from the air in the early morning. This process does not occur in the lab or greenhouse, and this system cannot be studied indoors in a realistic manner.

Laboratory experiments are usually conducted under conditions that are simplified and controlled by design. Even if you are able to set up the experiment and answer the question that you posed, you cannot know how well it depicts similar processes in nature. The solution to this problem is to link lab and field studies. They each can provide unique information but also have unique limitations. Field observations and experiments should be followed by lab studies to learn more about the ecological mechanisms that could cause the field result. Lab studies should be followed by field studies and "natural experiments" to learn whether the lab results are realistic and whether they hold at larger spatial and temporal scales (Diamond 1986).

Organizing a Season of Experiments

Organizing a field season requires thinking of both the big picture (e.g., your questions) and the day-to-day activities. Plan out your season in your field notebook, not just in your head. You might be tempted to begin by focusing on your methods, such as treatments and sample sizes, but resist this urge. Start instead by writing down the questions you want to answer that season. Make sure you have a firm handle on these questions and that they are answerable. It is amazing how often the manipulations that are conducted don't effectively and directly answer the question that we want to address. This "disconnect" can often be avoided by being very explicit about the question and then asking whether the manipulation really is the most appropriate match for the question. Consider whether alternative approaches could be informative, effective, and more efficient.

Before you begin your experiments, troubleshoot them as much as you can. What will happen if your organisms are harder to find than you expect? What will you do if you get bad weather? Come up with contingency plans. Try to anticipate all the potential things that could go wrong and how you will respond to them. Run some of these potential problems and solutions by someone who knows the system (your major professor or a colleague) to make sure that your ideas are on track.

Once you start to do your project, let your observations direct your next steps. Go where your system wants to take you. Constantly reassess what you can do and what alternative explanations and paths could be. Take stock of your progress and your preliminary results at least once a week. Go in the directions that you determine will be the most profitable and don't feel constrained to follow the plan that you set for yourself before you started. As you redesign your experiments and also when you write them up, develop your best story (your most interesting results) and not necessarily the question that initially got you into the system. You may be interested in the historical development of your thoughts, but your audience is likely to want to hear your most exciting results placed in the most interesting framework, whether you came upon that framework deliberately or by dumb luck. A lot of the best scientists are both very lucky and smart enough to recognize that they have stumbled upon something unexpected and novel. Keep your eyes open and be willing to take off in a slightly different direction.

At the end of each field season, come to grips with your results and what they mean. We recommend that you do this in two ways. First, analyze your data as soon as you can. In the field, often this analysis is only a summary of your treatment means and variances so that you get a qualitative picture of the results that you are finding. Even without access to a computer, you can and should get a feel for your data. Back at home, a complete statistical analysis lets you determine which effects are real, which experiments you want to repeat, and which new directions you want to think about. Second, soon after returning home, write up your results as a preliminary paper. This has multiple benefits. If you do not write your methods during your field season, you are very likely to forget key details later. In addition, the reading that is involved with assembling the introduction and discussion can be particularly valuable. This process lets you put your questions and your results in a broader perspective and gets you familiar with the pertinent literature. (Note to perfectionists: Don't get hung up here. It is more important to get a preliminary draft of your results than to do a complete literature review.) You often get new ideas for questions that you want to ask next by thinking about the bigger ecological issues and by seeing what other people have found in related studies. This process also forces you to reevaluate and reprioritize your questions and plans for future research in light of what you have just found. Finally, writing up your results each season gives you first drafts of your publications.

CHAPTER 4

Analyzing Patterns and Data

Hypothesis Testing and Statistics

The first step in doing research is to have a clear question or hypothesis in your mind. If you are vaguely interested in a system (an organism or an interaction), you are not ready to do experiments. You must be able to formulate your ideas into a clear question. Without a clear question, there is no end to the data (relevant or otherwise) that you may feel compelled to collect. In addition, it is not acceptable or efficient to collect data and then fish around for low p-values. For these reasons, you should walk the fine line between having a clear and focused question and keeping your eyes open for unexpected answers and new ways to rephrase your question.

It is better to start with a well-formulated question than to start by running experiments. A clear question will stimulate relevant experimental manipulations and statistical analyses. For example, you might suspect that species A (elephants) reduces the population size of species B (ants). Your testable working hypothesis is

that the population of ants will be lower in the presence of elephants than in the absence of elephants.
Because you have a clear hypothesis in mind, you are
now ready to design an experiment. You can conduct a
manipulation, removing elephants from half of your
plots and keeping the other half as unmanipulated
controls. You can measure populations of ants in these
two treatments and compare the difference that you
measure against the difference that would be expected
by chance. Based upon a statistical analysis, you determine the likelihood that your null hypothesis that elephants don't reduce ant populations can be rejected.

Statistics allow you to evaluate whether the differences
caused by your treatments are likely to be real differences or are likely to reflect random noise. You can
find older studies without proper controls, or replication, or even statistics. However, it is virtually impossible
to get experimental studies published nowadays without statistical analyses.

We often confuse statistical significance (indicated
by a probability level) with biological significance (indicated by the size of the effect). For example, we might
hypothesize that diet affects the body size of rodents.
If we feed two groups different diets, we can be fairly
certain that the two will not grow to the same size. What
we really want to know is whether they grow to sizes
that are different enough to consistently observe and
to be worth caring about. Very small differences can be
statistically significant but not produce consequences

that are important. In other words, we want to know if effects are "biologically significant," although we often use "statistically significant" as a proxy.

To remedy this confusion, we must consider both statistical and biological significance (that is, p-values and effect sizes) when we evaluate our experiments and when we present them (see chapter 6). It is insufficient to report that two populations were different and give a probability value ($p < 0.05$ or $p = 0.023$) or to report that differences were not significant (ns). Along with any significance test, report the effect size. This can be done by showing a picture (perhaps a histogram with the means and standard errors for each of the populations) or by reporting that ants were 35% more numerous when elephants were absent. Box 3 shows how to calculate effect size. When interpreting your own results and those in the literature, separate statistical significance from biological significance.

Although statistical tests are absolutely essential for manipulative ecology to progress as a science, the emphasis in our field on significance tests may be a bit overly zealous (Yoccuz 1991). By convention, we have decided that we will consider two populations to be different if the probability of their being the same is less than 0.05. Before we start, we know for certain that no two populations are exactly identical. Finding that our estimates are significantly different at the magical 0.05 threshold depends on how variable the populations are and on our sample size. If the sample size is very large, two populations can be statistically different with means that are quite close. On the other hand, if the sample size is small, two populations that are actually quite different will not appear significantly different at the 0.05 level. We find it baffling that a group of intelligent and thoughtful people can enslave themselves to this essentially arbitrary number.

Ecologists know that no two populations are identical just as no two people are. When we test a null hypothesis that two populations are the same, we are not calculating the probability that they are truly identical but rather the probability that we can detect a difference between them. Statistical significance is really a property of the organisms, the data, and the experimenter's ability to make distinctions. Unfortunately, the 0.05 threshold has become an absolute wall between "real" results and "negative" results. Why should we be allowed to say that two populations are truly different if $p = 0.049$ but not be allowed to say much of anything if $p = 0.051$? In both cases our inference about the populations being different will be wrong approximately 5 times out of 100. We should be aware of the arbitrary nature of the 0.05 threshold and interpret results accordingly. Whenever possible, we should also give the calculated p-value rather than reporting p greater or less than 0.05. If $p = 0.001$ we can be more confident that the result was not caused by chance than if $p = 0.05$. Similarly, our confidence about $p = 0.06$ should be different than about $p = 0.60$.

Box 3. *How to calculate effect size*

The effect size is a measure of the magnitude of experimental effects. Consider a simple experiment in an African savanna in which large herbivorous mammals are excluded experimentally from some large-scale plots but natural densities of herbivorous mammals are allowed in others (the controls). Such an experiment tells us about the consequences of extinctions of large mammals on populations of grasshoppers. Since grasshoppers eat mostly the same food resources as the mammalian herbivores in this African savanna, their densities were higher in plots without mammals than in those with mammalian herbivores (see figure and table). We can report that this difference was statistically significant ($p = 0.014$), although without the figure, this by itself doesn't give us any idea of the size of the effect that mammals had on grasshoppers. One way to describe the size of the effect is to report the absolute difference in grasshoppers between the two treatments:

$$|\text{mean}_{\text{exclosures}} - \text{mean}_{\text{controls}}| = \text{absolute difference}$$

In this case, the absolute difference in mean grasshopper numbers per sample was $|15.00 - 6.97| = 8.03$. This absolute difference between treatments is not as useful as comparing the change in means caused by the treatment relative to the control:

$$|\text{mean}_{\text{exclosures}} - \text{mean}_{\text{controls}}| / \text{mean}_{\text{controls}}$$

$$= \text{relative difference}$$

In this case, removing mammalian herbivores increased grasshopper numbers by 115%. This gives us the sense that mammalian herbivores had a huge impact on grasshopper abundance. (We could conceivably report that controls with mammalian herbivores had 54% fewer grass-hoppers than experimental plots, although this generally makes less sense than reporting differences relative to the unmanipulated controls as we did in the first case.) A third useful technique is to report standardized effect sizes scaled by a measure of the variance or noise involved with measuring them. This is often done by calculating the difference between the means divided by their pooled standard deviation (data in table):

$$\frac{|\text{mean}_{\text{exclosures}} - \text{mean}_{\text{controls}}|}{\text{standard deviation}_{\text{exclosures + controls}}}$$

For grasshoppers, this would be: $(15.00 - 6.97) / 4.89 = 1.64$. Such effect sizes are unit-less, which allows us to compare the effects found in different studies and using different response variables. Again we see that this is a relatively huge effect.

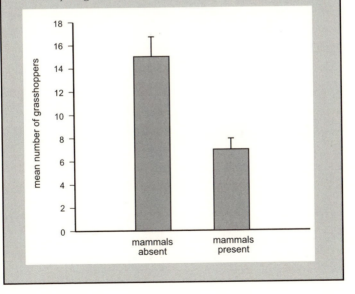

Box 3. *Continued*

Experimental Parameter	Value
Exclosure (no mammals)	
mean number of grasshoppers	15.00
standard error	1.71
sample size	3
Control (with mammals)	
mean number of grasshoppers	6.97
standard error	0.94
sample size	3
Both treatments (exclosure & control)	
mean number of grasshoppers	10.99
pooled standard deviation	4.89
sample size	6

Alternative hypotheses

Scientists have been urged to design null hypotheses that can be critically tested and rejected (Popper 1959, Platt 1964). This is sound advice, although it doesn't work very well for many ecological questions (Quinn and Dunham 1983). Many ecological hypotheses are not simply true or false. For example, suppose we are interested in understanding the role of competition in structuring communities. This is not a question that can be falsified by conducting a simple experiment. Evidence for competition can be found in most systems. However, so can other processes like predation and parasitism, disturbance, and so on. Instead of rejecting hypotheses and asking yes/no questions, we should be weighing alternatives and asking: how important is competition relative to other processes? We should be more interested in discovering the size of the effect caused by competition rather than trying to reject a null hypothesis that it does or does not operate. Unlike some other scientific disciplines, ecological principles are not universal. Finding a single counterexample will make us rethink our working hypothesis about the force of gravity. However, finding a single counterexample does not disprove our ideas about competition. Similarly, unlike other disciplines, ecological hypotheses are rarely mutually exclusive. Even if competition is found to be important, predation may also be important.

Ecologists should strive to explicitly develop alternative hypotheses to explain the patterns that they observe (Platt 1964). By developing a set of alternatives, you don't become emotionally attached to the hypothesis that you selected initially. Rick's son Jesse built a crude leprechaun trap in kindergarten, baited it with "gold" rocks, and placed it out on the eve of St. Patrick's Day. The next morning Jesse checked the trap. Unbeknownst to him, one of the rocks had dropped into the grass. He was immensely excited by this evidence that the leprechaun had visited, taken one of the pieces of gold, and gotten away. The next year, he built a more sophisticated trap. Although leprechauns failed to visit in this second year, he still felt confident of their existence based on the previous season's results. It is unfortunate that Jesse did not consider alternative hypotheses for the rock's disappearance. It's worth wondering how many leprechauns we each catch in our research careers.

Too often ecologists have become advocates of a polarized point of view (e.g., density dependence versus density independence) and then spent their careers tenaciously defending it. We are all under pressure to produce significant results. If you start with a list of alternative hypotheses, you can reduce this stress because you are much more likely to turn up something interesting. In addition, if you propose several alternatives, any or all of which may be valid, you are likely to be objective in testing each of these alternatives. Often after you have tested a hypothetical ecological mechanism, it becomes clear that there are alternative mechanisms that could also have contributed. It is generally better to have thought of these alternatives before conducting your experiment rather than dealing with them after the fact. However, even after the fact, it is much better to test alternative explanations for your results than to ignore the possibility of their existence.

Testing alternative hypotheses is a very efficient way to assure that you have something to say when you get done. If you become enamored of a particular process, you may conduct experiments to test whether this process is important in your system or to elucidate the

Box 4. *Generating alternative hypotheses*

Once you have identified a pattern that is interesting to you, think about a working hypothesis to explain or produce that pattern. Next consider alternative hypotheses that could also produce that pattern. Try the following list of possible factors as alternatives that could also have produced your pattern:

☐ abiotic factors (precipitation, temperature, light, fire regime, etc.)

☐ predators, parasites, and disease

☐ mating factors (sexual selection, nest-site availability, opportunities for offspring, etc.)

☐ microhabitats (shelters from abiotic conditions, predators, etc.)

☐ disturbance attributed to human influences or natural causes

☐ genetic or ontogenetic (developmental) influences

Your list of alternatives can get long and unwieldy, but this is an important step in doing good science. You don't necessarily have to test all of your alternatives, although getting them all down on paper for consideration is a first step. Prioritize them based on how compelling and how testable each one is, and work through them in that order.

mechanisms by which this process operates. Following this strategy of focusing on only a single hypothesis, you'll have a story to tell only if your results come out one particular way. If, instead, you consider alternative hypotheses, you have something to talk about no matter what you find. Box 4 provides suggestions for generating alternative hypotheses in ecology. In addition, Kevin Rice recommends avoiding the "house of cards" research program. If all of the secondary hypotheses that you are interested in testing require a particular outcome to be true in your initial hypothesis, then you put yourself under too much pressure to demonstrate your primary hypothesis, whether or not it is actually true.

Answering yes/no questions will often take the form of rejecting hypotheses. But we have argued that many ecological hypotheses cannot be rejected in this way. Instead, it may be more useful to devise a list of alternative hypotheses, acknowledge that most or all of these working hypotheses may be valid, and then attempt to determine the relative importance of each of the alternatives. This process is akin to partitioning the variance in ANOVA that is due to each of the working hypotheses. For instance, Rick observed that spittlebugs, plume moth caterpillars, and thrips all fed on seaside daisy along the California coast. He wanted to know how those three herbivores affected each other. Instead of just testing the hypothesis that they competed, he examined the relative importance of interspecific competition, predators and parasites, and plant genotype on the success of each of these common herbivores (Karban 1989). This was done by including all three factors (competition, predation, host plant effects) in one experiment and partitioning the variation in herbivore performance (survival, fecundity, etc.) that could be attributed to each factor.

This experiment considered three different factors that could affect herbivore performance, but it only examined the effects of complete removal of two of those factors, competitors and predators. In other words, it compared the effects of the complete removal of competitors (or predators) with natural levels of competitors (or predators). In the jargon of statistics, there were only two levels, all or none, and each of these levels was replicated 30 times. This design works the best if the effects of the predictor (in this case presence/ absence of competitors) on the response variable are linear. Unfortunately, ecological effects are often not linear. Examining the natural relationship between numbers of competitors and performance over space or time can provide valuable intuition. When you suspect that the relationship between the predictor variable that you are manipulating and the response variable may be nonlinear, a useful design involves many levels of the predictor variable, and each level need not be replicated. For example, you could assign many different levels of competitors so that the selected levels spanned the entire range of values that you observed in nature. This design would be analyzed with a regression rather than an ANOVA. Regression is similar to ANOVA except that it has many levels (rather than two or a small number in ANOVA) and it does not require replication at each level, as in ANOVA. This regression design does not assume a linear relationship between variables, and even lets you determine the shape of the relationship. However, regression with many levels works best with one or a few predictor variables but becomes difficult and unwieldy when multiple factors are examined simultaneously. In addition, it would have been difficult to set up experimental treatments with many levels of predators or competitors. Other treatments, such as fertilizer application, lend themselves more readily to experiments with many different levels.

Recently biologists have become interested in Bayesian statistics, in part because they allow us to evaluate how well multiple working hypotheses fit data (Hilborn and Mangel 1997, Gotelli and Ellison 2004). Instead of rejecting a null hypothesis, the result of a Bayesian analysis is an index of confidence in each of several hypotheses. A Bayesian approach lends itself beautifully to evaluating alternatives and has the potential to be a valuable tool for ecologists capable of using it. Unfortunately, Bayesian analyses require more computational sophistication than the methods that field biologists have traditionally used, and "ecological detectives" without a lot of background and confidence in mathematics will probably find them inaccessible at this point.

Negative Results

Earlier we discussed the statistical procedure that lets you reject null hypotheses with varying levels of confidence. When we fail to reject a null hypothesis (i.e., $p > 0.05$), does that mean that the null hypothesis is true? In other words, if we ask whether two populations are different and we find that we cannot conclude that they are different at the 0.05 level, should we conclude that they really are the same? The answer to both questions is no. Based on the information we have, all we can conclude is that we failed to find the difference or effect we had hypothesized. Our statistical tests give us far more power to reject hypotheses than to accept negative results as reality. In most cases we have very weak power to evaluate whether two populations are similar. Furthermore, we rarely use relevant statistics to address this issue. As such, many negative results (by which we mean results that are not statistically significant) in ecology never get published, a loss to the scientist who did the work and to the ecological community that never got to hear about it.

Techniques are available that allow you to evaluate whether the effect caused by one factor is as great as the effect caused by another factor. Using these techniques, negative results that fail to reject hypotheses can become as informative as "positive" results. Unfortunately, larger sample sizes are required to reliably infer that a treatment had no effect on a population than to conclude that it did have an effect (Cohen 1988). Our ability to accept a negative result will also depend on the effect size that we are looking for. The heart of this issue revolves around the expected effect size, the degree to which the treatment means were different (Cohen 1988). By convention, statisticians tend to define a small effect as a difference of 0.10 (10%) or less and a large effect size as 0.40 (40%) or greater (Cohen 1988). If we expect a really large effect of our treatment (see box 3), we can be more confident that one truly didn't exist than if we were only expecting a small effect. Cohen (1988) provides a very readable discussion with worked examples of how to calculate the probability that your result of "no significant difference" reflects the actual biological situation at your study site. Many statistical packages calculate a value for power that tells you how likely you were to find a significant difference given your amount of replication and your observed effect size (whether small, large, or in between).

It is often informative to compare a factor that didn't show a significant treatment effect to one that did. You can then calculate the probability that the negative effect (that you failed to find) was not as large as an effect or result that you found to be statistically significant. For example, Rick tested the hypothesis that early herbivory that made wild cotton plants more resistant to later herbivores would also increase growth of the plants (figure 4, Karban 1993). He failed to find this effect of induced resistance on plant growth. In fact, early damage tended to reduce plant growth, although these effects weren't statistically significant. Had he missed effects that were real? He could be reasonably sure (99% confident; all confidences calculated using techniques in Cohen 1988) that he hadn't missed a large effect (that the means were 40% different from equal) but much

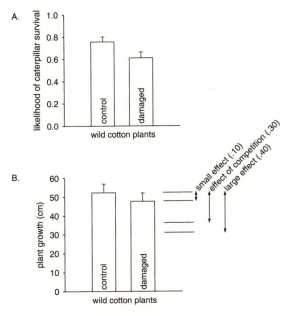

FIGURE 4. Effect of early herbivory in wild cotton plants on (A) caterpillar success and (B) plant growth (Karban 1993). Early damage by herbivores ("damaged") significantly reduced the likelihood that later caterpillars would survive (A). However, the study failed to find a significant effect of early herbivory on plant growth (B). In a simultaneous study, plant competition significantly reduced plant growth by 30% (histogram not shown). The lines to the right of histogram (B) show the effects of early damage that would be expected based on a hypothetical large effect size (0.40 difference in means) or small effect size (0.10 difference in means), and by an effect size equal to that empirically found to be caused by plant competition. These comparisons allow us to compare the effect size that we found for early damage with a defined standard (Cohen's (1988) large and small effects) and with effects of other factors (plant competition) actually found in this study. For example, this examination allows us to report, "Early damage did not produce a large effect on plant growth and was likely less important than plant competition."

less sure (only about 20% confident) that he hadn't missed a small effect (that the means were 10% different). These values are a little difficult to put in perspective. In the same experiment, he found that intraspecific plant competition reduced plant growth. By comparing the size of the effect due to plant competition with that due to induced resistance, he could be 30% confident that effects of induced resistance were not as great as those of plant competition. Calculations such as these allow us to report both "positive" and "negative" results that are not statistically significant and to compare the relative importance of different effects. They should receive more attention from ecologists, especially those who work in systems where large sample sizes are possible.

A useful way of comparing effect sizes from different studies is *meta-analysis* (Gurevitch and Hedges 2001). Each study becomes one independent measure of the effect of a particular factor or response variable. For example, we might be interested in the effect of removing top predators on lower trophic levels in a variety of published studies. A meta-analysis lets us formally and quantitatively compare the results of many studies, and lets us put our own experimental results in a much broader context. For each published study, we can calculate an estimate of the effect size by comparing the means of the treatment groups (with and without top predators), standardized by the variance. (See box 3 for more on how to calculate effect sizes.) The meta-analysis can be used to evaluate whether conclusions from one study are generalizable. It can also provide information about the conditions under which extrapolation from our experimental results is warranted. For example, meta-analysis revealed that studies of trophic cascades (removal of top predators) produced larger effects in aquatic systems than in terrestrial systems (Shurin et al. 2002). Clearly this general conclusion would not have been possible based on the results of any single experiment. Meta-analysis can only be meaningful if it includes all available information on a particular question; this is another reason why publishing negative results can be critically important to advancing the field.

Path Analysis

Earlier we were critical of attempts to draw causal connections based on observations and correlations (chapter 2). We argued that manipulative experiments were more powerful tools to establish cause and effect in ecology. The frustrating truth is that manipulative experiments are not always possible. Ethical concerns keep us from conducting experiments on rare species. Ethical and logistical concerns prevent us from doing many of the large-scale experiments that would be most informative. Manipulative experiments take time; sometimes policy makers want ecological information now (or yesterday). One alternative method that can be used to examine correlated factors is path analysis.

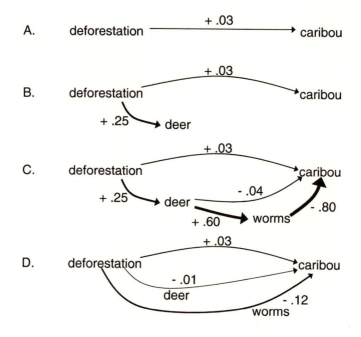

FIGURE 5. A hypothetical study (with pretend numbers) showing how path analysis can be used to identify important indirect effects of deforestation on caribou (see figure 1 for the natural history of this system). If it were possible to measure the direct effects of deforestation on caribou populations, it might be found that clearing the trees made more food available to caribou, producing a slightly positive effect (A). However, deforestation also would benefit populations of white-tailed deer, allowing them to invade areas where they had been previously extremely rare (B). Deer have a small negative competitive effect on caribou and a much larger indirect negative effect that is mediated by an increase in meningeal worms that devastate caribou populations (C). The net effects of each path can be calculated by multiplying the components. Thus the effect of defoliation on caribou mediated by worms that are more abundant because deer are now more abundant is calculated as

Path analysis is a technique that allows a researcher to try different causal schemes (paths) and to evaluate which of those paths does the best job of explaining the observed patterns (Shipley 2000, Mitchell 2001). The diagrams with arrows that connected weather, plant stress, and herbivore numbers in chapter 2 are different path diagrams that represent different hypotheses of cause and effect. The easiest way to evaluate paths is by using partial regression coefficients to estimate the strength of individual paths (arrows in the diagram) and goodness-of-fit statistics to estimate the overall model (see Mitchell 2001 for a very clear description with worked examples). Path analysis can shed light on the likely validity of several different cause-and-effect hypotheses. It is particularly useful for appreciating indirect effects (figure 5). However, results from path analysis are sensitive to several assumptions (Shipley 2000,

---

$(+.25)(+.60)$ $(-.80) = (-.12)$. When the effects of these three paths are calculated, the defoliation has a positive direct effect on caribou $(+.03)$, a negative small effect mediated by competition for resources with deer $(-.01)$, and a larger negative effect mediated by a shared disease $(-.12)$ (D). It would be possible to observe this net effect of deforestation on caribou populations without path analysis, although path analysis can help suggest the ecological mechanism that caused the net effect. It suggests that an experiment to remove the worms should eliminate the negative effect of deer observed on caribou. Path analysis can also be used to simultaneously examine other ecological effects, such as shared predators (Schmitz and Nudds 1994, de Castro and Bolker 2005), to determine the relative importance of each.

Mitchell 2001). Sometimes path analysis has been found to provide considerable insight into ecological processes, although other times it has led to conclusions that were not supported by experimental data (Petratis et al. 1996, Smith et al. 1997). More complicated path analysis based on structural equation modeling rather than regression may solve many of these problems, although it is more difficult to use (Shipley 2000).

In the end, path analysis is based on correlations. We include it here because it's relatively easy to use and it can provide information when experiments aren't possible. It can be used for evaluating observational data to estimate effect sizes and prioritize alternative hypotheses to decide which are likely to be important. It can be especially useful when you have numerous different possible players and are unable to manipulate all of them. Path analysis can be extremely valuable early in a study before committing to time-consuming manipulations. However, at present, conclusions drawn from path analysis cannot be as firm as those inferred from manipulative experiments.

CHAPTER 5

Working with Other People

When we aspired to be biologists, we imagined that the scientific process was totally objective; the truth could be separated from lesser hypotheses in a manner that was removed from social interactions. The longer we stay in this business, the more we are struck by the reverse. Science is a social endeavor, most ecologists are human beings, and successful workers are able to convince others of the value of their ideas.

In this chapter we will consider some of the social situations that you are likely to encounter as a graduate student and as a professional ecologist. Graduate training takes place in universities, institutions that were established during the Middle Ages and that cling unconsciously to medieval European conventions. These include what Damrosch (1995:18) described as "the indentured servitude of graduate student apprentices and postdoctoral journeymen." Perhaps because universities originated from monastic traditions, a high degree of zealous dedication and self-discipline is expected. Any less is likely to be met with disapproval. Many administrators at universities are starting to talk about "work-life balance," but the academics that you encounter (e.g., your major professor, search committees, etc.) may have no such balance in mind.

When you are picking a graduate program, pick a professor rather than a university. The prestige of the university granting your degree is far less important that the level of intellectual, emotional, and financial support that you receive along the way. You should try to get a support package that will keep you from needing outside employment, but you probably shouldn't pick a program based on a few thousand dollars more or less. Grad school is a mistake if your goal is to get rich; you can increase your income by doing just about anything else. But grad school may be a great place to be if you are passionate about ecology.

Make sure that you can communicate well with your major professor. Visit potential professors before accepting any offer. Ask about what has happened to previous students who have moved on. How many have finished their degrees? If they didn't, why? What jobs did they get? Correspond with them. Would you like to be in their shoes 5 or 10 years down the road? Professors have track records, and you can expect to face many of the same situations and eventual successes as past students. If you are not yet a grad student, you will probably underestimate the importance of the relationship with your major professor. However, it will color everything that happens to you, so choose carefully.

One subject that grad students and major professors often butt heads about is authorship. Each subdiscipline has a slightly different tradition about what constitutes authorship, with lab-intensive programs including the major professor as an author more often than field-oriented programs. It's a good idea to ask former and current students about what it is like to publish jointly with your potential future major professor. The Ecological Society of America has developed a policy statement for authorship (www.esa.org/aboutesa/governance/codeofethics.php). To be considered an author, all participants must make substantial intellectual contributions. Simply providing funding does not confer authorship privileges. Even if it is awkward, discuss authorship (including author order) before designing a study. You may want to keep this fluid, but having a discussion early on can prevent big disappointments later.

Involve your committee members so that they can provide advice and support much as your major professor does. Choose committee members who are going to give you the most help. Get to know them by taking their seminars, attending their lab meetings, talking to them about your project. Try to use the same people for your thesis committee, exams, and so forth, so that you know each other well. The better they know you and the more invested they feel in you, the more help they are likely to give you when it comes time to submit your manuscripts and apply for jobs. Don't hesitate to run your career ideas, research proposals, grants, manuscripts, and job materials past your committee members.

If they are too busy they'll tell you, but at least they will know that you are trying hard. Don't make the mistake of "getting lost" interacting only with your own lab.

The switch from undergrad to grad school can be a shock. For one thing, the currency has changed. Doing well as an undergrad involved taking classes and getting good grades. Once in grad school, grades become virtually obsolete. Unless you are making the switch from a master's to a doctoral program, nobody cares about your classes and grades. The next step is all about research, and the currency of research is publications. Prioritize the things that will result in publications. When allocating your work time, think long and hard about those activities that won't result in pubs— do them only if they will make you better over the long term. As a rule of thumb, until you make significant progress on your first research project (that is, have a rough first draft of preliminary results), you probably shouldn't take on additional activities, because you may be using them to avoid your demons.

As a grad student or as a professional ecologist, you can often expand the scope of your research by collaborating with other scientists. Good collaborations, like other mutualistic interactions, generally involve organisms who can offer skills or expertise that their partners cannot easily acquire on their own. Perhaps one person is a good chemist or mathematician and another has little expertise in these areas but has a lot of knowledge of natural history. Together this team is capable of going places that neither person could go alone. Sometimes people who have similar interests but different personalities are able to work together very effectively; an idea person who has trouble finishing projects can work well with a pragmatist who is a little less creative. The main disadvantage of collaborating is that you lose some control over the content and pace of the work. One person may be in more of a hurry than the other and this can be frustrating to both. Despite this constraint, the advantages of collaborating are great, and the rugged individualist who works completely independently is becoming less common in our field. Thirty years ago, most papers in *Ecology* were single-authored; today very few are.

Another valuable way to interact with colleagues is by attending meetings. Even if you find these events stressful, it is worth making yourself go and trying to interact as best you can. Be kind to yourself at meetings by not having unreasonable expectations and by pampering yourself. (For example, take a nap or go for a walk if you are feeling fried.) Don't feel like you need to introduce yourself to the big dogs in your field. Any and all contacts can be beneficial.

Meetings give you the opportunity to let other people know what you are doing, to find out what other people are up to, to get feedback, and most importantly, to schmooze with a few people. Since personal contacts are so important, getting familiar with the people in your field is extremely worthwhile and will help you start collaborations, get your manuscripts and grants accepted, and let you feel part of a community.

Research projects and careers in ecology put you in contact with a wide diversity of people including administrative secretaries, reserve stewards, and resource managers, among others. Cultivating good relationships with these people can be mutually rewarding. Their help can often determine the success or failure of your project. Even if you are shy or unsure, make an effort to communicate your respect and appreciation for these facilitators.

One way to get research accomplished is by hiring helpers. They can do repetitive work, freeing you to do more creative tasks. Research assistants can also get things done at your field site when you can't be there. In addition, some projects require a lot of hands, and assistants can allow you to answer questions that you could not address as one person. However, there can be severe downsides to hiring helpers. They have much less invested in the quality of the work than you do. For many assistants, it's just a job. Assistants generally don't have the expertise and intuition for the system that you do. Often when we have asked other people to do routine tasks, we are surprised when the tasks aren't done "right." There are so many little things, "tricks" if you will, that each of us takes for granted so that it is very difficult to convey all of these to another person no matter how detailed the instructions. All of this subtle intuition comes from working with your organisms. If you hire someone else to do the hands-on work while you do the paperwork, you will miss out on developing intuition about the details of your system (and maybe about the big questions as well). One way to minimize these risks is to work alongside your helpers. That way you can provide more quality control and also develop critical intuition. Helpers are expensive; hiring them generally involves writing grants, progress reports, and endless other details, and this paperwork can make you the administrator and your assistants the biologists.

In summary, science is a far more social endeavor than we had imagined. Effective communication with people around you is an important skill well worth cultivating. In the next chapter, we offer our suggestions on communicating what you find with ecologists with whom you don't directly interact on a daily basis.

CHAPTER 6

Communicating What You Find

Communicating is an essential part of doing field biology, although it requires very different skills than scientific investigation. Learning about nature is fun, but the field of ecology only advances when you communicate what you have learned. We have never been able to make a lick of sense of the argument that a tree in the forest hasn't really fallen if nobody is there to hear it. However, from society's point of view, if you don't make other interested people aware of what you have learned, then essentially you haven't learned anything.

Not all attempts to communicate are successful, and this aspect of ecology has an enormous effect on whether your findings and ideas will have an impact. In the sixth and final edition of *The Origin of Species,* Charles Darwin included "an historical sketch of the progress of opinion on the origin of species." Essentially, Darwin explained why his ideas really were different from those of numerous predecessors who, by 1889, wanted some of the credit and fame for the theory that Darwin had expounded. Most of the authors were easy to deal with; they had simply missed the main points of the theory of natural selection. However, one author was more troublesome for Darwin, and he wrote, "In 1831 Mr. Patrick Matthew published his work on 'Naval Timber and Arboriculture', in which he gives precisely the same view on the origin of species as that (presently to be alluded to) propounded by Mr. Wallace and myself in the 'Linnean Journal', and as that enlarged in the present volume. Unfortunately the view was given by Mr. Matthew very briefly in scattered passages in an Appendix to a work on a different subject, so that it remained unnoticed. . . ." Matthew understood the principles and their significance, but he didn't effectively communicate what he had grasped. He had the same impact as if he had never had the ideas in the first place.

This problem is not limited to Victorian times. For example, MacArthur and Wilson (1963, 1967) revolutionized evolutionary ecology with their theory of island biogeography. Unfortunately, years before MacArthur and Wilson, Eugene Munroe proposed the same equilibrium theory, along with empirical support for the species-area relationship for butterflies in the West Indies, and detailed models to explain it (Munroe 1948 [his thesis], Munroe 1953 [an obscure proceedings]). Munroe did little to communicate what he had found, and the scientific community remained unaware of his insights (Brown and Lomolino 1989). These examples illustrate that it matters where you publish. Make sure that you are reaching the largest and most appropriate audience. Both Matthew and Munroe are forgotten footnotes in the history of ecology because other, better communicators independently came to similar conclusions. How many Matthews and Munroes have there been whose potentially revolutionary advances have never been repeated and communicated?

Publications are the currency of our field. Some journals are more influential than others and reach a much wider audience. Rating systems for journals have been developed to measure this influence. These ratings fluctuate and are available at several web sites, including www.sciencegateway.org/impact. However, because these rankings are created for all sciences combined, they are cumbersome and difficult for ecologists to use. The best ways to get a sense of the respected journals in your subdiscipline are to ask several experienced ecologists and to read a lot yourself.

Writing and talking about your work communicates your ideas and findings. In addition, the act of organizing your work by presenting talks and writing papers facilitates understanding what you know, what you don't know, and how the various pieces fit together. As almost any seasoned teacher will tell you from painful experience, we often think that we have a pretty good grasp of a subject we are about to lecture or write about. However, once we sit down and look for the actual words that we are going to use, we realize that we haven't thought through the ideas. The act of writing or speaking clarifies your thoughts and will probably be valuable for you, independent of the value of communication.

Many ecologists who communicate their work successfully use an outline to organize talks and papers. The outline can be either formal with roman numerals (Rick's preference) or informal with a bulleted list (Mikaela's preference). When we're organizing our work, we use a list when we aren't sure about how to order the various ideas. Then we give each of the things we've written down a number or color code that helps to group the ideas that are similar or related. Next we figure out which of these should go first and how to connect them to make a logical argument. If you think you don't like outlines, but you haven't actually used them to write a professional paper, we recommend you give them another try. They may seem like a waste of your time, but they make you more efficient in the long run and help you write a more organized paper.

It is a good idea to write up your results after each field season. The preliminary literature review that this requires helps give you a sense of how your experiments fit into the bigger picture. Writing up your results will also make it clear what you have nailed down conclusively, and what parts of your argument are weak and need further testing. Coming to grips with what you have will also help you design the next steps. Doing this may seem like extra work. It's not. You wind up using much of this preliminary draft when you write the paper up for real. It also makes writing up the final paper a much less daunting task. Finally, it is much easier for colleagues or committee members to appreciate what you have and provide helpful feedback if they can read a manuscript rather than if you tell them something vague about what you think you found. In the sections that follow, we will offer some suggestions for organizing your work into (1) a journal article, (2) an oral presentation, (3) a poster, and (4) a grant proposal.

Journal Articles

Journal articles are the bread and butter of biologists. Writing papers can seem daunting at first, but as you begin to recognize the formula for writing them, they will become easier. Journal articles serve the important function of archiving what you have learned and making it available to share with the rest of the ecological community.

Most of your paper should be written in the past tense. You are describing what you did, what you found when you did it, and how you interpreted those findings. Some authors write the introduction in the present tense, describing the current state of knowledge.

Expect to revise your manuscript. Rick sometimes likes to think of his first draft as a place to get his ideas organized and out of his head. He tries to approach his second draft much more critically, from a reader's perspective. Is the story clear? Does the logic follow?

Does his writing express exactly what he was thinking? Sometimes he tries to imagine that his father is reading the manuscript. His father had no formal training and therefore didn't have the jargon and preconceptions of a trained ecologist. Rick asks: Would he follow what I am saying? How would I change my paper so that he would understand it?

Inexperienced writers sometimes imagine that they should sit down and write a polished version. Instead, it may be helpful to think of the writing process as four distinct steps (Lertzman 1995). The first step is figuring out what you have to say. The second step is organizing your thoughts (with an outline or other technique that works for you) so that you can present a logical argument. The third step is putting down the words that make your argument. Read this draft critically, imagining how it will sound to your audience. Then, as a fourth step, carefully craft your writing so that it makes your points concisely and convincingly. Separating these tasks may help you get started. Including all four, if you don't already, should improve your writing.

Most journal articles are expected to follow a standard format: abstract, introduction, methods, results, discussion, and conclusion. (Even articles in *Science* and *Nature* are written in this format, although in a manner that is less easy to spot.) In the following sections we give you some information that will help you see the formula you should follow when writing papers.

Title and Abstract

The paper starts with a title and an abstract (although we find it is easier to write these last when we have a good sense of the main points and their significance). The title tells what the paper is going to be about. Don Strong, the editor of *Ecology*, says it should present the main result rather than just including the key words. For example, "Fire increases butterfly diversity in riparian and woodland habitats" is a more informative title than "A study of the effects of fire on butterfly diversity in two habitats."

The abstract provides a summary of the paper. As such, the abstract should include a sentence or two of rationale, the main results, and what they mean. The abstract has to be concise and clear. Far more people will read your abstract than other parts of your paper. Even if they do read the entire paper, reviewers and critics of all kinds will make their decisions about the paper and your story based largely on the title and abstract.

Introduction

Your introduction should present your question and explain why it is interesting. Your first paragraph or first few sentences should set the stage for your question. How do we (other ecologists) think about this subject? One effective way to begin your paper is by stating a problem or observation that everyone agrees is impor-

tant and grabs our attention. If it is not obvious why
your problem is important, then you must make a case
for why we should care enough to read your paper.
Could solving your problem shed light on a bigger
issue, for instance? The makeup of your audience determines how you should frame your question so that
they find it interesting.

We find an introduction that poses a general question to be much more effective than one that simply
describes an organism, study system, or topic (although
the latter is how many students are tempted to begin).
So for instance, don't tell us that you are interested in
wooly bear caterpillars. Instead lead with the question—
tell us what is known generally about how herbivores
choose food and then describe how your study will add
to this knowledge by considering food choices of wooly
bears. If your audience won't immediately find this
question highly relevant to their interests, explain why
choice of host plants is critical to understanding other
ecological and evolutionary questions. In other words,
make sure that you have explained why we as ecologists
need the information you are providing. For example,
don't just assert that it is critical to know the rates of
parasitism in lemmings. Instead explain why knowing
about parasitism could help us understand why populations cycle. Sometime later in the introduction or
methods, tell us the natural history of your system. An
introduction that begins with a general "hook" will catch

the attention of more of your audience than one that starts with wooly bear caterpillars or lemmings as interesting organisms.

Setting the stage for your question puts it into a broader context; this often includes general statements about the current wisdom on your subject. You may want to reference work that has preceded your paper. This may include work on your system, or on other systems that asked similar questions. This review should only be included if it helps explain where you are coming from, that is, why you want to ask this particular question. References should not be included just to show that you are familiar with the literature.

Your introduction should contain an explicit statement of your question—tell your audience exactly what you are asking. What hypothesis or hypotheses are you testing? (Your hypothesis is an explanation of your question, although it doesn't have to be [and probably shouldn't be] stated as a null hypothesis.) If you have a topic but haven't put it into the form of a question, you should. Doing this little exercise can also clarify your thoughts.

We like to end our introduction by giving either a formal listing of the questions that we will answer (for a paper) or a brief answer to the question that we posed at the beginning (for a talk, see next section). This solution to the problem lets the audience see where we are going to go in our methods and results. Also, because this formula is relatively standard, some readers will skip to the last paragraph of your introduction to decide whether the questions you plan to ask are interesting enough to keep them reading.

Methods

In your paper, your methods must be described clearly enough so that the work could be repeated by someone else. This should include a description of where, when, and how you applied your treatments and took your measurements. Make sure that the reader understands the motivation for each experimental procedure. So instead of just launching into the details, start the description of each experiment with something on the order of, "To test the hypothesis that wooly bear caterpillars choose the most common host plants. . . ." At the end of each methods subsection include a brief description of the statistical analyses that were used.

Your methods section should be kept brief because methods are often boring and because reviewers, editors, and readers look unfavorably on papers with lengthy methods. Graduate students often assume that they need to explain every detail of their projects, but this is surely not true. Only include information that is directly relevant to the story that you want to tell. For example, perhaps you kept detailed data each day on temperature, percent sunlight, or precipitation because you thought that it might help explain variation in crawdad feeding events. However, you didn't find any relationship, and your story evolved in other directions. We don't need to hear about these details even though you may be tempted to show readers how thorough you have been. Information that isn't immediately relevant will only clutter up your presentation and obscure your story.

Methods sections often include descriptions of the natural history of your system. (This can also go near the end of your introduction, preceding the list of questions that you will answer.) Tell us enough, but only enough, natural history so that we can follow the important points of your experiments and interpretation.

Results

The results section tells what you have found—your data and statistical analyses. In general you will not be permitted to make a statement if you don't have statistical analyses to support it. Results should be described in the way that tells your story logically. Often students describe their results chronologically, in the order in which the experiments were conducted. Instead of this chronological ordering, a topical organization that answers your main questions in a logical progression is generally more effective. We like to organize our methods and results sections by questions or experiments. We use subheadings as titles for each question, and these are repeated in order in both sections so that the reader can easily connect a method with the corresponding result.

Results should be reported in the text or summarized as figures or tables. Generally, use only one of these three for any given result, although a figure should be interpreted in the text without repeating the information that is clearly visible. Text or figures are more effective at conveying most kinds of information than are tables. Text is the default choice; if you can convey a result effectively by describing what you found, do so. Figures are particularly good at conveying relationships between factors, although actual values are generally obscured. Tables show exact values but are not good at presenting relationships between factors. As a simple rule of thumb, the most informative way to present your results may be to follow these two steps. First, use a figure to show the data: for example, use a histogram to show that female salmon produced approximately 60 eggs in low-selenium streams and 20 eggs in high-selenium streams. Second, report the percent change in the text: "High selenium levels dramatically reduced salmon egg production. Female salmon produced only about one-third as many eggs in high-selenium streams as in low-selenium streams [statistical test comparing these means can go here]."

When you are describing your results, emphasize what you learned and not your tables and figures. The tables and figures merely illustrate what you are describing in the text. For instance, "Adults consumed 40% more than juveniles (table 1)" is preferable to "Table 1 shows the consumption rates of adults and juveniles."

Similarly, when citing published work, describe the results and not what the authors did: "Males were bigger than females (Brown 2000)" rather than "Brown (2000) reported that the sexes differed in size."

Simple figures and tables are better than more complicated ones. Figures and tables should be described clearly with captions and titles, so that someone looking at them can make sense of the results without necessarily reading the whole paper. Your audience must be able to discern what you actually measured. This is often achieved by clearly labeling the axes of your figures. The fewer the number of treatments presented in one figure, the easier it is to grasp. Don't combine figures unless viewing all the information at one time adds meaning. It often helps to include a legend identifying the treatments directly on the figure. Letters, numbers, and lines must be readable. As a rule of thumb, when preparing figures, use a font size of at least 20 and a line width of .015 inch for borders, histogram bar edges, error bars, etc., so that the figure is legible when it is reduced for publication.

The most commonly used figures are histograms (bar graphs) and then scatter plots. When you use histograms, it is easier for a reader to grasp the meaning if you use fewer bars. Under most circumstances, error bars should be presented with all histograms. These are important because they give the reader a sense of the noise around the signal. Standard errors are used most commonly and show the noise or precision around your estimate of the mean. The standard deviation is used when you want to show the amount of variation *per se*. Occasionally error bars make a figure so busy so that the signal becomes unrecognizable, and only under these circumstances should error bars be omitted. Scatter plots are also commonly used by ecologists. The line that describes the best fitting model can be added to the scatter plot when the model is found to be significant.

Tables should be used only when repetitive data are essential to tell your story. For many arguments, fewer data are more effective than more. Only include those variables that are relevant. Don't use tables (or any part of the results section) as a core dump for your field notebook. One common application for tables is to summarize statistical tests. For example, in an analysis of variance, the sums of squares, degrees of freedom, F-ratio, and p-values all provide unique information. If this information is not required to make a convincing case, then include just the F-ratio, degrees of freedom, and p-value in parentheses in the text.

Two suggestions will help you communicate more effectively to a biological audience. First, your results should be presented in biology-speak rather than statistics-speak. This highlights the biological results and not the statistical tests. For instance, tell us, "males were twice the size of females (student's $t = x$, df $= y$, $P = 0.0z$)" rather than "the student's t-test with y degrees of freedom showed a statistically significant effect at the 0.0z level of gender on body size." Second, always present the effect size, not just the level of statistical significance—"males were twice as big as females" rather than "males were significantly bigger." The effect size tells us about the biological relevance of the result (see box 3), whereas the statistical significance tells us how likely it is that the result was caused by chance.

As we mentioned in the section on methods, it may be tempting to include all of the experiments and observations that you performed. Don't do it. Include only those results that are connected logically—that tell your story. Variables and effects that are not relevant to the story should be omitted, or the audience will be distracted from the main points. Many authors make the mistake of trying to include all the data they have instead of thinking about what pieces are needed to tell the best single story.

Discussion

You should explain what your work means in the discussion. To do this, first restate concisely the most important result. Then interpret it. How do you make sense of what you found? What evidence have other studies brought to bear on the question? Do your results resolve the question that you posed in the introduction? Then, as they become relevant to your story, add in the other results of your study, and interpret them. Often the results of experiments will suggest subsequent hypotheses. These should be described in your discussion. You may be able to generalize from your results as well as those of others. Do any useful models or paradigms result from this work?

Believe your results and interpret them in that manner. If you didn't find what you were expecting, don't excuse your results and talk about how they might have been different with a larger sample size, or if you had controlled for other factors, or if you did the work in a different place. If you don't believe your results yourself, don't take our time telling us about them. If the evidence supports your hypothesis but you still don't believe your results because you feel unconfident and doubtful about everything, talk about this with a therapist, but don't let these doubts pervade your presentation. Also remember that most discoveries are surprises. If you already know the answer, then the question isn't particularly interesting.

Throughout your paper you should tell a cohesive story. Don't wander from your central point. Rather, your presentation should present a tightly reasoned argument that is evident from start to finish.

Conclusion

Papers should end with conclusions (although these are often missing). The conclusion, like the abstract, is a concise summary of your results and their significance. End with a sentence or two that states the important consequences of the findings. Leave us with the take-home message. Don't have too many take-home messages. Most papers have only a single real lesson. Make sure that it doesn't get lost but instead is painfully obvious for those people who will read only the conclusion.

It may be useful to once again state and answer the question that you posed at the start. Don't trail off with some weak non-conclusion like "this is a good system" or "more work should be done." Of course more work should be done following every study. Instead leave us with what you have learned. If we remember anything from this work, what should it be?

Box 5 provides a summary and checklist of our suggestions about writing journal articles.

Oral Presentations

Hearing a talk is a very different experience for the audience than reading a paper. As a speaker, you should be aware of the differences so that you can use them to your advantage. Interacting with a person is far more compelling than reading a book. Think about how likely you are to put down a book without finishing it compared to how likely you are to walk out of a play or movie. The more you involve your audience, the more successful you will be at holding their attention and having them remember what you say. As a result, you definitely do not want to read your talk; reading is far more difficult to listen to than is speaking directly to an audience. A conversational tone is easier to absorb

Box 5. *Journal article checklist*

Title

☐ Does the title summarize the main result?

Abstract

☐ Does the abstract tell your story?

Introduction

☐ Does the first sentence or two of your introduction "hook" the reader by setting the stage for the question your manuscript answers?

☐ Do you explain and justify your question(s) rather than just extol the virtues of your study organism?

☐ Have you briefly summarized previous work that informs your current question? (Briefly—this should not be an exhaustive literature review.)

☐ Did you end your introduction by clearly listing the questions your manuscript addresses?

Methods

☐ Do you briefly explain the relevant natural history of your organisms and/or study system? (Some of this information may fit more appropriately in your introduction.)

☐ Have you described your methods thoroughly enough that another ecologist could repeat your experiment but briefly enough that space-pressured journals won't send your manuscript back?

☐ Have you started the description of each experimental method with a phrase justifying why it was done?

Box 5. *Continued*

☐ Did you include a brief explanation of each statistical procedure you used?

☐ Are each of your methods relevant to your overall story?

Results

☐ Are your results presented in a logical order to help your reader follow your story (not in the order in which you did your experiments, if that is different)?

☐ Have each of your results been presented only once (in the text, a figure, or a table)?

☐ Does your text inform your readers of your results as much as possible instead of simply referring them to your figures or tables?

☐ Are your results described in biological rather than statistical terms?

☐ Have you presented effect sizes for each of your results?

☐ Are each of your results presented in terms of your overall story?

Discussion

☐ Did you restate your main results briefly and interpret them?

☐ Did you generalize to larger ecological concepts where appropriate?

☐ Does the information in your discussion relate to your initial questions? Does your story seem cohesive?

Conclusion

☐ Did you hit your reader over the head one last time with your take-home message?

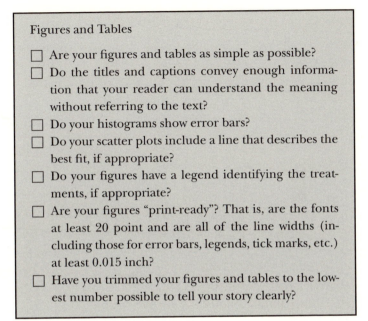

Figures and Tables

☐ Are your figures and tables as simple as possible?

☐ Do the titles and captions convey enough information that your reader can understand the meaning without referring to the text?

☐ Do your histograms show error bars?

☐ Do your scatter plots include a line that describes the best fit, if appropriate?

☐ Do your figures have a legend identifying the treatments, if appropriate?

☐ Are your figures "print-ready"? That is, are the fonts at least 20 point and are all of the line widths (including those for error bars, legends, tick marks, etc.) at least 0.015 inch?

☐ Have you trimmed your figures and tables to the lowest number possible to tell your story clearly?

than a speech. If possible you should give your talk without notes, but this is less critical than not reading it. If you are worried about forgetting your talk, use an outline or let your slides lead you. If there are specific critical concepts or points that you are prone to forget, link them to a particular slide. When you get to that slide (usually a picture), it is your cue to remember to provide a particular piece of your story.

When you are giving a talk, look at the people in the audience; eye contact will help to involve them. A common mistake that speakers make is to talk to their slides; instead, talk to your audience. Leave the room as bright as you can so that you can see the faces of the audience and they can see yours. It is a biological fact that dark rooms and slides with dark backgrounds put people to sleep; don't use them. A lot of light is much more important than having photographs that show up really well on the screen. Which would you prefer, having some large fraction of the audience dozing while really sharp pictures go up or having the audience alert and attentive at the expense of some photographic quality?

Stay close to your audience. This lets you relate to them more effectively. If the podium is too far away, move it closer or don't use it. Come out from behind it and address the audience directly. Walk around a bit in the space you are given as well. It is amazing how much the simple act of walking from one side of the screen across to the other helps keep your audience alert and focused.

Tailor your talk to your particular audience and keep the information clear. Tell people things that they have the background to relate to. Imagine what the interests of your audience are likely to be and prepare your talk with that bias in mind. Avoid jargon (abbreviations, specialized words, measures, techniques) that is not familiar to almost everyone in your audience. Once people tune out or get lost, it is hard to get them back. You cannot expect people to grasp equations or complicated theories that they have not come to grips with previously. When you come upon new or complicated written material, you can read it slowly, digest it, and reread it until you understand it. There is little chance for this to happen in a talk, so don't lose your audience by including this kind of material. It is fine to include a theoretical motivation for your work or to develop a new theoretical argument, but use words rather than equations if you are presenting a talk, and be extra careful that you are clear and the material is accessible. If you are presenting a talk that is mostly theory, you may have no choice but to include an equation or two. Spend a lot of time with any equation, explaining in words and perhaps simple graphs what each of the clusters of terms means to a biologist.

Good talks require as much structure and planning as good papers. Don't imagine that you can just wing your talk off the top. When you are organizing your talk, build it around your take-home message. Figure out what your punchline is, set it up right from the start, tell the audience what it is going to be, deliver it, and then remind the audience what it was during the conclusion. A paper requires elaborate documentation of the existing literature, of your detailed methods, of the statistical analyses, and so on. These are generally boring to listen to and should be minimized in a talk. Your talk must have one, single coherent story to tell. Don't be tempted to include all the loose ends—just your best single story. This is true for a paper but much more important for a talk. Stress the concepts and not the details. Try not to "switch gears" (that is, tell two stories) during your talk; your audience won't take much away from such a mixture. If you feel that you don't have enough material for one coherent story, then spend some time thinking about how to organize it so that it seems like a single story and then more time "smoothing" the transition between the pieces. One way to do this is to present a single question (or short series of interrelated questions) in the introduction and one punchline that integrates all of the parts of your story in the conclusion.

We like to use signposts to let the audience see the structure of the talk and how each piece fits into the bigger picture. When you're reading a paper, you rely on the subheadings and paragraph indentations to recognize transitions. The audience listening to a talk needs similar signposts. Rick often likes to make a slide showing the outline of his talk or to put his outline on the blackboard if one is available. Then he refers to the outline at various times during the talk. (Remember though that the blackboard will not be visible if the light is dim.) It helps people follow his train of thought if they can see where he is going throughout the presentation. Mikaela uses a higher tech approach. She puts up an outline slide early in her talk and repeats it at various times throughout her talk, highlighting her current position to show the audience where she is at that point.

Let the questions dictate the outline. Don't use the traditional methods, results, and discussion that you would in a paper. Instead, integrate these three for each

question or subquestion that you address. For each experiment, tell us in a sentence the question that was
asked. Describe the approach you used in a sentence
or two and then the result in a sentence or two. What
does this result mean, and did it stimulate the next
question? Then repeat this sequence for the next experiment. Make sure that the audience follows your
logical path.

The methods in a talk should be very abbreviated
and contain only what is absolutely essential for you
to tell your story. Although a talk should not attempt
to provide the listener with the ability to repeat the
experiment, many speakers make the mistake of including far more methods than are necessary or interesting. Very abbreviated methods are often effectively
integrated with results in talks.

When presenting your results, aim for simplicity.
Again, this applies to papers as well, but simplicity is
critical for a talk. Pictures are better than words or tables.
Make sure that you explain the axes of graphs. The audience isn't already familiar with the variables on each
axis, and needs to identify these before the data you
are presenting can be appreciated. Keep the graphics
simple. Three dimensions and a bunch of colors rarely
help tell the story. Never show a table that your audience cannot read. In general, tables are much less effective in talks than in papers and should be used spar-
ingly, if at all. Tables used in talks must be simple, with
large, easily read characters.

Similarly, each slide should make only a single point. Never put more on a slide than can be presented easily. Similarly, don't put up slides containing writing that you merely read. Slides with words on them should be as abbreviated as possible—a few words or phrases but not sentences. Don't put more than ten or fifteen words on a slide, and don't use a font size smaller than 24.

Let your slides guide your story, but don't make them your focus. Plan to give your talk as if you didn't have slides. (In case there is a technical failure, you should always be able to give your talk without using slides at all.) They should be the background that illustrates what you are saying. Don't structure your talk so that it is a progression of, "This slide shows this and this next slide shows that." You'll give a better talk if your slides illustrate the story rather than if they are the whole story. Reading the exact words from a slide is less effective than sharing the spontaneity of putting the words together to express your thoughts.

This is not to say that you should not practice your talk. The more you practice it (especially in front of a real audience) the better it will be. If a real audience is not always available, saying the actual words out loud is much better than thinking them. Mikaela used to think it was too embarrassing to practice a talk out loud with her housemates overhearing her. Then she learned it was more embarrassing to give a poorly practiced talk. Tape recording a practice (or actual talk) will help you improve and also will help you learn your talk quickly.

If Rick has to give a talk that he really cares about or if he is strapped for time, he records a practice and then listens to it (or at least has the tape on) several times while he's doing other things. It is amazing how much this helps.

Show only as many slides as your audience can digest. As a rule of thumb, use about one slide per minute of talk. Many speakers make the mistake of including far too many slides and needing to rush at the end or annoy the audience by going overtime. To avoid this, time yourself each time you practice. Also, we like to pause after we have made an important point. This provides emphasis and gives listeners a chance to make sure they absorb the last message. When you first begin giving oral presentations, your nervousness is likely to cause you to speak faster than you practiced. Concentrating on well-timed pauses will help you to pace your talk.

Spend extra time preparing your introduction and conclusion. These are the only parts that some of the audience will hear. Everyone is most alert at the start, so tell them what questions you are going to ask and what you found. Even if you lose some of them at some point, they will have heard your punchline. Similarly, if they haven't followed all of the talk, the conclusion should provide all of the take-home message. Students are often impressed by several well-known, older scientists at Davis who nod conspicuously during the talk but always seem to ask good questions afterwards. When this occurs, a lot of the credit should go to the speaker for highlighting the important points before the lights went out and then again after they came back on.

A talk from a person who is slightly nervous is often better than a talk from someone who lacks sufficient adrenaline. However, excessive nerves can make a talk difficult to follow. Convert your nervous energy into larger, more flowing motions rather than small, repetitive jitters. Practicing your talk can increase your confidence (for some people this means one time, for others it means six full run-throughs). Also remember that the more times you give talks, the easier they become. The experience of giving talks will get better, although we know this offers little consolation for right now. As much as possible, turn your anxiety into enthusiasm. A mistake many beginning speakers make is to be self-deprecating and apologetic. Replace this with enthusiasm; your feelings become contagious.

One of the best parts of a talk is the question session at the end, although this can be the most intimidating. We like to leave a lot of time at the end for questions (10–15 minutes for an hour-long talk and 2–5 minutes for a 12–15-minute talk). We already know what we have to say, but we're excited to hear the spins that other people will place on our results. Often new and exciting ideas come up in the questions after a talk. We sometimes ask a friend to take notes during the questions so that we don't have to remember all the suggestions. If the room is large or the questioner soft-spoken, repeat the question for the audience before answering it. Make sure you understand a question before responding to it. It is fine to paraphrase the question and ask the questioner if you have it right. It is also fine to say that you don't know the answer to a question. You are permitted to say that the point raised is an interesting one and you will think more about it or design an experiment in the future to test it. You might ask if the questioner can think of a way to test the notion that he or she is bringing up. Try hard not to be defensive. Students who fail oral exams are generally not the least prepared but rather the ones who become defensive and argue with their committees. Very occasionally you are confronted by a jerk who is aggressive about asking questions and won't give up the floor until you acknowledge his point (it almost always is a he). One way to handle this situation is to say that you would like to move on but you would be happy to talk more with him after the seminar is over.

Box 6 provides a summary and checklist of our suggestions about talks.

Posters

Posters are much more similar to talks than they are to papers. However, most posters suffer from being prepared like manuscripts. Remember that people at meetings are burnt out. Do you enjoy reading a lot of fine print when you are viewing posters? We never do.

Box 6. *Oral presentation checklist*

Note: Please see also the "Journal article checklist" for reminders of good communication habits in ecology.

Introduction

☐ Did you structure your introduction around your take-home message?

☐ Did you eliminate most of the citations and other details you would include in a manuscript to help keep your audience's attention focused?

☐ Did you end your introduction by showing a slide that clearly indicates the questions you will address?

Methods, Results, and Discussion

☐ Have you integrated your methods, results, and discussion in a way that makes your story easy to follow (even if that means doing several series of methods, results, discussion; methods, results, discussion)?

☐ Have you minimized (that is, practically eliminated) your methods? (You should be prepared to discuss them if asked, but assume that people will not be interested in them during your talk.)

☐ Did you explain how the results of your first experiment generated your next question and experiment, so that your audience understands your thinking?

Conclusion

☐ Did you leave your audience with one take-home message?

Figures and Tables

☐ Did you show your results in pictures and figures rather than just describing them?

- [ ] Did you minimize or eliminate the use of tables, which are hard to grasp during a talk?
- [ ] When presenting your figures, have you planned to indicate to your audience what the x- and y-axes are?
- [ ] Have you kept your graphics simple, so that they make only one single point, have a large font size (24 point or larger), and include very few words (10–15 maximum)?

General Structure and Presentation

- [ ] Did you figure out how to present your information as one single, coherent story to help your audience follow you?
- [ ] Did you review your whole talk to include signposts so that your audience follows the structure you have created?
- [ ] Have you practiced your talk (especially the introduction and conclusion) until you are absolutely comfortable with the information in it?
- [ ] Are you able to give your talk without reading it? (You can rely on your slides or your outline to remind you what to say next.)
- [ ] Have you practiced making eye contact with your audience (instead of with your slides), moving about the room enough to keep your audience engaged?
- [ ] Have you carefully examined your talk for jargon you might not even realize you are using?
- [ ] If your talk includes an equation, have you planned how you will make it readily accessible to your audience?
- [ ] Have you created roughly one slide for each minute of your talk?
- [ ] Have you timed yourself to make sure your talk does not go overtime?
- [ ] Are you prepared to give your talk without any slides at all in case of a technical problem?

Instead, we want the take-home message in a simple, readily available package. Posters are to scientific communication what *USA Today* is to journalism. You should present only the headlines and the briefest explanation to make your point. Everyone who walks by your poster should immediately know your question and know the answer. Those people who are interested can ask you to explain in more detail and can read the paper when it comes out.

Your introduction should be no more than a few sentences stating the conventional wisdom or explaining the justification for your question. Next, present your results as pictures (figures and photographs) that tell your story. Move logically from one result to the next, making sure not to include more information than your viewer can easily and quickly digest. You should either skip the methods completely or include only enough to make your results meaningful. The details of your experimental design, sample sizes, and so on should not be included. After each result, you can include one sentence of "discussion" that makes each result more general or relates it to your big question. At the end of your poster you might want to include a sentence or two that explicitly answers the question that you posed at the start. Similarly, a sentence or two (but no more) explaining the significance of your results, how they fit into the big picture, is a useful way to conclude.

One advantage of presenting a poster is that you can walk interested people through your story. This is more

Box 7. *Poster checklist*

Note: Please see also the "Journal article checklist" for reminders of good communication habits in ecology. Because your poster will rely heavily on pictures and figures (as opposed to words), you may especially want to refer to the "Figures and Tables" section.

Title

☐ Does the title summarize the main result?

Introduction

☐ Did you limit the introduction of your question to one or two sentences?
☐ Did you clearly present the question(s) your poster will answer?

Methods

☐ Is your methods section extremely brief?

Results and Discussion

☐ Are your results presented mainly as pictures (histograms, scatter plots, etc.)? Have you shown the differences in treatments with photographs where appropriate?
☐ Do you briefly explain the significance of each result?
☐ Are each of your results presented in terms of your overall story?

Conclusion

☐ Did you include a sentence or two briefly answering the question(s) you posed at the start?

General

☐ Does your poster contain only the headlines?

effective than asking them to read the thing. In addition, they have the opportunity to ask you questions about things they don't understand or suggest other experiments and directions. Thus it can be a good use of your time to hang out with your poster and interact with viewers as much as you can.

In summary, the poster that we have described contains less than one-tenth the number of words of most posters at ecology meetings. It tells only a single story and does this using only "headlines." It contains no or few references and no methodological details. It has figures and photographs but rarely tables. Details and statistical analyses are not included. It is much more effective at conveying information than the poster that is essentially a manuscript pasted on a board.

Box 7 presents a summary and checklist of our suggestions about posters.

Grants and Research Proposals

Selling Your Research Ideas

The purpose of grants and research proposals is to sell your plans about work you want to do. You want your committee to agree to give you a degree if you fulfill the objectives in your research proposal, and you want people to give you money in response to your grant proposal. In addition, your proposal provides two less obvious functions: it forces you to develop a research plan and it forces other people to consider your ideas more carefully than they might otherwise and give you feedback. Grant and research proposals involve more salesmanship than research talks or papers. Therefore, a slightly different emphasis is required. As you prepare a proposal, focus on three things: (1) novelty, (2) clarity, and (3) feasibility.

Your proposal outlines what you want to do. First, it must be exciting. You must justify how your work will forward your subdiscipline or the way people apply science to solve problems. Obviously not every proposal is going to change the way all scientists think, but those people who work on your question or on your system should be influenced by your work. If it is not clear to you how this will happen, think long and hard about how to justify your work in these terms. Emphasize this justification throughout your proposal. If justifying your proposal sounds too vague, think about answering questions such as: What makes your proposed work significant? What is the value of the work? If all goes well, how might other people be able to use your results? How will other people inside and outside of your field view the contribution of your work? The biggest mistake that students make writing proposals is not including enough justification.

Second, your proposal must be simple and clear, even more so than a scientific paper. Reviewers often get many proposals to read at one time, and all reviewers have better things to do than read them. Unlike the scientific papers they review, these proposals are not necessarily about subjects that the reviewers are already interested in or knowledgeable about. From the reviewer's first glance at your proposal, you have only a few seconds to convince him or her to read further and pay attention while reading. Then you have only a few minutes to convince the reviewer that your proposal is worthy of funding from a budget that can in many cases fund only 10% of the proposals in the stack. If your writing is not clear, the reviewer may not do the work required to figure out what you are trying to say. The proposal must be convincing to the meticulous reader as well as the rapid skimmer. A well-known colleague who serves on many NSF panels calls this the two-glasses-of-wine problem. He does all of his reviewing at the end of the day after two glasses of wine at dinner. Successful proposals must be clear enough to make sense to him under those conditions.

Third, you must convince readers that your proposal is feasible. Nobody is going to give you money or assurances of a degree unless they are convinced that you can complete the work you propose and that your work will answer the interesting questions that you have posed. There is an inherent contradiction in this process, since your grant must appear both feasible and also exciting, which means it must be novel and break new ground. You must convince people simultaneously that your ideas are important and that the experiments you propose can be accomplished. The best way to convince people that you can pull off your experiments is to use techniques that have been used before (include citations to previous literature). It is even better to be able to say that the techniques are old hat for you or people in your lab. An excellent way to show that your plan is feasible is to present preliminary data. For a research proposal this often involves doing a first year of fieldwork that addresses the main question. This heavy emphasis on preliminary results frequently means that researchers propose work that they have largely completed and use the money to generate the next set of preliminary data.

Organizing Your Proposal

The organization of your proposal differs slightly from that used for talks and papers and is generally less canalized into a specific style. Different universities and granting agencies require different sections, and it is important to know these requirements and to fulfill them. Below we will consider the form and content of a proposal that would apply generally for many graduate groups and funding agencies. Many proposals include (1) an abstract or project summary, (2) an introduction, (3) explicit objectives, (4) experiments, justification, and interpretation that correspond to each objective, and (5) a budget. Other sections that are often helpful include separate discussions of the significance of the experiments, the potential pitfalls associated with the experiments and your solutions, a timetable for completion of each experiment, and a justification of the

budget. You can get more detailed advice about preparing proposals in a recent book by Friedland and
Folt (2000).

The abstract or project summary shares many simi-
larities with that discussed earlier for a journal article.
It comes first, although we write it last. It needs to be
crystal clear and capture the excitement and rigor of
the proposal. It should describe the big-picture problem that you are addressing. Next, emphasize a justification for your work and explain its significance. Describe what you expect to find and explain why your
results will be influential in your field. The project
summary generally presents fewer results than abstracts
for papers but can contain a few sentences about your
approach.

The introduction to the proposal must get the reader
excited about your questions. Justify why this work is
important. This is difficult. Even after having given this
advice to grad students, more often than not we find
their proposals could use more justification. How does
your work relate to the big questions in ecology and
why should we care? Frame your work in terms of the
questions that you will address rather than systems that
you will use. This advice holds even if you chose your
project because you were interested in the system. In
fact, it holds especially if you chose your project because of your system. You shouldn't write "This question is important" in your proposal. Instead, explain
why it is important. For example, if you are studying

tick population dynamics, say, "Tick-borne diseases cause *x*-number of diseases per year. To better control tick-borne diseases, we must first understand the population dynamics of ticks," etc.

Your introduction should explain what has been done to date to motivate your question. It is often best to first present the general question and then describe how your specific research on a particular system will address that broader question. Start general and get specific. Here you can also tell us about the natural history of your system, but only if this information is immediately useful in understanding how you will answer your question.

Next present your objectives. These can be long term (more than you can accomplish now to address a big-picture question) and short term (the actual goals of experiments in this proposal). The objectives should be presented explicitly and numbered. With each objective you can include the hypothesis that is tested and a rationale for that objective. Using Christopher Columbus as an example, van Kammen (1987) differentiates objectives, justification, and hypotheses. If Columbus submitted a proposal to Queen Isabella and King Ferdinand, his *objective* would be to establish a new trade route to India to bring back three ships full of spices. He would *justify* his proposal by explaining that a water route to the west could be faster and less expensive than currently used routes, and such a route would increase their wealth and international power.

By fulfilling these objectives he would test the scientific *hypothesis* that the earth was round. He would further justify his proposal by attempting to convince Isabella and Ferdinand that it was feasible to accomplish these objectives and that he, Columbus, had the necessary know-how and experience.

Each objective should be addressed by specific experiments. It is often useful to number these experiments exactly as you have numbered the objectives. A rationale and experimental design should be presented for each experiment. Describe how you would do each experiment and demonstrate that you can accomplish each procedure. Finally describe how you plan to analyze the data from each experiment.

Tell us how your results will be interpreted: "If experiment 1 gives this result, I will conclude the following." Interpretation of the results may or may not be in its own section in the proposal. Remember that hypotheses in ecology must be testable although they are not necessarily falsifiable or mutually exclusive. At this point you might want to include another section or paragraph entitled "Significance" if the importance of your work has not been extensively discussed and stressed.

We like to include a small section detailing potential pitfalls. This section provides damage control and troubleshooting. Anticipate questions that the reviewers are likely to have and address them here. Try to explain how you will turn apparent misfortune into a situation in which you learn a lot. Describe here how you will interpret different outcomes from your experiments. The best projects are those that give interesting results no matter what the outcome. Try to design your experiments so that you are not dependent on getting one particular result to have something interesting to say. If you have designed a research program that will let you gain new and useful perspectives about nature no matter what the outcome, make certain that you stress this feature.

We also like to include a timetable for our objectives and experiments. This helps establish that we have thought about how and when we will get everything done. A timetable helps make the work appear feasible. The timetable is also very useful to refer to when doing the work.

Box 8 summarizes our suggestions about grants and research proposals.

Three things about the granting process should be kept in mind. (1) Grants are competitive and it often takes several attempts before a grant gets funded. Don't get discouraged. (2) At the same time, take the comments to heart. We find it helps us get our emotions under control if we put the comments aside for a few days after getting negative criticism. It can be frustrating to get comments that seem to miss the point. If a reviewer missed our point, it indicates that we need to rewrite the proposal so that even a sleepy idiot with two glasses of wine in his or her stomach can follow our

Box 8. *Grant and research proposal checklist*

Note: Please see also the "Journal article checklist" for reminders of good communication habits in ecology.

General

☐ Is your proposal novel and exciting, and have you made this very clear to your reader?

☐ Have you made clear the value to the larger ecological community of your proposed work?

☐ Is your proposal simple and clear, easy enough for an exhausted non-ecologist to understand at the end of a long day?

☐ Is your proposal feasible, and have you explained its feasibility in a way that will be convincing to your reviewers? When possible, have you proposed to use established techniques?

Project summary/abstract

☐ Does your project summary capture the excitement of your proposed research?

Introduction

☐ Have you taken extreme pains to justify your proposed work?

Objectives

☐ Have you stated each of your objectives explicitly?

☐ Have you justified each of your objectives?

☐ Have you written hypotheses that address your objectives?

☐ Have you designed and described experiments that address your hypotheses and objectives?

logic. Almost invariably, the comments will contain extremely useful suggestions as well as these few misconceptions. (3) Don't let the granting process dictate what your questions are. A funded grant signifies the approval of our peers and administrators. In addition, some projects require money to pursue. But many ecological questions can be answered for relatively little money. The granting process is a very conservative one. You can only get funding for ideas that everyone is already comfortable with. This retards innovation. We have seen over and over again how a few dollars can get graduate students and senior faculty alike to change their research priorities and pursue projects that were not necessarily burning questions for them. Our advice is to follow your own intuition. Proposals are approved by a committee of scientists. Why give up anything as personally important as your research direction to an anonymous committee? Would you let a committee of scientific peers approve your choice of a partner over the next three to five years?

Hard work often determines productivity, and productivity often determines success. Pick the questions that are most exciting to you whether you get funding or not and you are more likely to work hard enough to be successful.

CHAPTER 7

Conclusions

There is a card game called Mao that is popular on several university campuses. One of the rules of Mao is that players cannot ask or explain the rules. New players join the game and must deduce the rules by observation and trial and error. A player who fails to follow a rule is given a penalty.

Doing field biology can be a lot like playing Mao. The rules of field biology often go unstated. In this handbook we have attempted to make the unstated basic rules of the game explicit. You may or may not wish to follow the rules, but you might as well know what they are because you will face the consequences for choosing not to follow them. Unfortunately life and ecology are both complicated, and there are *also* long-term consequences for following the rules too literally. Below we highlight some of the rules of our game, as well as some of the potential costs of following them too assiduously:

Rule 1. Manipulative experiments are a powerful and highly respected technique to establish cause-and-effect relationships in ecology. Experiments lend credibility to your study.

Cost of Rule 1. Experiments are only as good as the intuition that went into the hypotheses being tested. Make sure you find time to know your organisms or your experiments won't teach you much. In other words, make time for observations and natural history.

Rule 2. Test clear hypotheses by using inferential statistics.

Cost of Rule 2. Don't get drawn into treating ecology as a science of truly falsifiable hypotheses and universal laws. Generate alternative hypotheses and evaluate the relative importance of each one.

Rule 3. Increase the statistical power of your experiments with a large sample size of randomly assigned, independent replicates.

Cost of Rule 3. Replication comes at the expense of scale and therefore of realism.

Rule 4. Plan your experiments carefully and evaluate your results frequently.

Cost of Rule 4. Don't get trapped insisting on answering your initial question. Keep working even when your questions and experiments aren't perfect. Be opportunistic, and pay attention to the directions in which your system is trying to send you.

Rule 5. Write proposals and apply for funding.

Cost of Rule 5. Unless you like administration, don't let writing proposals replace fieldwork for you. The granting process is very conservative, so do the projects that are the most exciting to you even if you don't get funded.

Rule 6. The currency for researchers (and grad students) is publications. If you're a grad student who still holds an undergrad mentality that grades and classes are useful currencies, realize that the rules have changed.

Cost of Rule 6. Just as grades never did perfectly mirror what you learned from classes, neither does a long list of grants and publications perfectly mirror learning about nature and advancing the field.

Unfortunately the rules reward short-term goals that may not be consistent with your longer-term goals. The good news is that most of us got into this business because we like being outdoors and learning about nature. You can and should make your job reflect your interests. You are likely to have more control of this as your career advances. While you play the game, keep your eye on the big prize: your own personal and professional priorities. This is your life! You will be more successful if you're enjoying it.

Acknowledgments

In this handbook we have gathered the advice of our teachers, role models, and colleagues together with our own personal experiences. Many people have shaped how we go about doing ecology, and we have borrowed heavily from what we have been taught formally and informally. We thank Anurag Agrawal, Winnie Anderson, Jim Archie, Leon Blaustein, Liz Constable, Will Davis, Teresa Dillinger, Hugh Dingle, Greg English-Loeb, Jeff Granett, Jessica Gurevitch, Henry Horn, David Hougen-Eitzman, Dan Janzen, Sharon Lawler, Rich Levine, Monte Lloyd, John Maron, Rob Page, Sanjay Pyare, Jim Quinn, Dave Reznick, Kevin Rice, Bob Ricklefs, Tom Scott, Jonathan Shurin, Andy Sih, Chris Simon, Sharon Strauss, Don Strong, Jennifer Thaler, Neil Willets, Louie Yang, and Truman Young, all of whom made valuable contributions to what's in this book. We are fairly certain that we have neglected to mention many others, and we apologize for these unintentional omissions. Thanks also to Apryl Huntzinger for support during the writing process. Finally, we have appreciated the help and encouragement of Robert Kirk, our editor at Princeton.

References

Augustine, D. J., and S. J. McNaughton. 2004. Regulation of shrub dynamics by native browsing ungulates on East African rangeland. *Journal of Applied Ecology* 41:45–58.

Baldwin, I. T. 1988. The alkaloidal responses of wild tobacco to real and simulated herbivory. *Oecologia* 77:378–381.

Bergerud, A. T., and W. E. Mercer. 1989. Caribou introductions in eastern North America. *Wildlife Society Bulletin* 17:111–120.

Brown, J. H., and M. V. Lomolino. 1989. Independent discovery of the equilibrium theory of island biogeography. *Ecology* 70: 1955–1957.

Cohen, J. 1988. *Statistical Power Analysis for the Behavioral Sciences.* 2nd ed. Lawrence Earlbaum, Hillsdale, NJ.

Crouse, D. T., L. B. Crowder, and H. Caswell. 1987. A stage-based population model for loggerhead sea turtles and implications for conservation. *Ecology* 68:1412–1423.

Damrosch, D. 1995. *We Scholars: Changing the Culture of the University.* Harvard University Press, Cambridge, MA.

Darwin, C. 1889. *The Origin of Species.* 6th ed. D. Appleton, New York, NY.

de Castro, F., and B. Bolker. 2005. Mechanisms of disease-induced extinction. *Ecology Letters* 8:117–126.

Diamond, J. 1986. Overview: Laboratory experiments, field experiments, and natural experiments. Pages 3–22 In J. Diamond and T. J. Case (eds.), *Community Ecology.* Harper and Row, New York, NY.

Felton, G. W., and H. Eichenseer. 1999. Herbivore saliva and its effects on plant defense against herbivores and pathogens.

Pages 19–36 in A. A. Agrawal, S. Tuzin, and E. Bent (eds.), *Induced Plant Defenses against Pathogens and Herbivores: Biochemistry, Ecology, and Agriculture*. American Phytopathological Society Press, St. Paul, MN.

Friedland, A. J., and C. L. Folt. 2000. *Writing Successful Science Proposals*. Yale University Press. New Haven, CT.

Futuyma, D. J. 1998. Wherefore and whither the naturalist? *American Naturalist* 151:1–6.

Gotelli, N. J., and A. M. Ellison. 2004. *A Primer of Ecological Statistics*. Sinauer, Sunderland, MA.

Gurevitch, J., and L. V. Hedges. 2001. Meta-analysis: Combining the results of independent experiments. Pages 347–369 in S. M. Scheiner and J. Gurevitch (eds.), *Design and Analysis of Ecological Experiments*, 2nd ed. Oxford Univ. Press, Oxford, UK.

Hilborn, R., and M. Mangel. 1997. The ecological detective: Confronting models with data. Princeton University Press. Princeton, NJ.

Holt, R. D. 1977. Predation, apparent competition and the structure of prey communities. *Theoretical Population Biology* 12:197–229.

Holt, R. D., and J. H. Lawton. 1994. The ecological consequences of shared natural enemies. *Annual Review of Ecology and Systematics* 25:495–520.

Huberty, A. F., and R. F. Denno. 2004. Plant water stress and its consequences for herbivorous insects: A new synthesis. *Ecology* 85:1383–1398.

Huntzinger, M. 2003. Effects of fire management practices on butterfly diversity in the forested western United States. *Biological Conservation* 113:1–12.

Huntzinger, M., and D. J. Augustine. n.d. Grasshoppers increase following removal of ungulate herbivores in an East African bushland. Unpublished ms.

Huntzinger, M., R. Karban, T. P. Young, and T. M. Palmer. 2004. Relaxation of induced indirect defenses of acacias following exclusion of mammalian herbivores. *Ecology* 85:609–614.

Hurlbert, S. H. 1984. Pseudoreplication and the design of ecological field experiments. *Ecological Monographs* 54:187–211.

Karban, R. 1983. Induced responses of cherry trees to periodical cicada oviposition. *Oecologia* 59:226–231.

Karban, R. 1987. Environmental conditions affecting the strength of induced resistance against mites in cotton. Oecologia 73: 414–419.

Karban, R. 1989. Community organization of *Erigeron glaucus* folivores: Effects of competition, predation, and host plant. *Ecology* 70:1028–1039.

Karban, R. 1993. Costs and benefits of induced resistance and plant density for a native shrub, *Gossypium thurberi. Ecology* 74:9–19.

Karban, R., and I. T. Baldwin. 1997. *Induced Responses to Herbivory.* University of Chicago Press, Chicago, IL.

Karban, R., and J. Maron. 2001. The fitness consequences of interspecific eavesdropping between plants. *Ecology* 83:1209–1213.

Kearns, C. A., and D. W. Inouye. 1993. *Techniques for Pollination Biologists.* University Press of Colorado, Niwot, CO.

Lertzman, K. 1995. Notes on writing papers and theses. *Bulletin of the Ecological Society of America* June 1995:86–90.

MacArthur, R. H., and E. O. Wilson. 1963. An equilibrium theory of insular zoogeography. *Evolution* 17:373–387.

MacArthur, R. H., and E. O. Wilson. 1967. *The Theory of Island Biogeography.* Princeton University Press, Princeton, NJ.

Maron, J. L., and S. Harrison. 1997. Spatial pattern formation in an insect host-parasitoid system. *Science* 278:1619–1621.

Marquis, R. J., and C. J. Whelan. 1995. Insectivorous birds increase growth of white oak through consumption of leaf-chewing insects. *Ecology* 75:2007–2014.

Mitchell, R. J. 2001. Path analysis: Pollination. Pages 217–234 in S. M. Scheiner and J. Gurevitch (eds.), *Design and Analysis of Ecological Experiments,* 2nd ed. Oxford Univ. Press, Oxford, UK.

Moore, P. D., and S. B. Chapman. 1986. *Methods in Plant Ecology,* 2nd ed. Blackwell Scientific Publications, Oxford, UK.

Munroe, E. G. 1948. The geographical distribution of butter-flies in the West Indies. Dissertation, Cornell University, Ithaca, NY.

Munroe, E. G. 1953. The size of island faunas. Pages 52–53 in *Proceedings of the Seventh Pacific Science Congress of the Pacific Science Association.* Volume IV: Zoology. Whitcome and Tombs, Aukland, New Zealand.

Newmark, W. D. 1995. Extinction of mammal populations in western North American national parks. *Conservation Biology* 9:512–526.

Newmark, W. D. 1996. Insularization of Tanzanian parks and the local extinction of large mammals. *Conservation Biology* 10:1549–1556.

Oksanen, L. 2001. Logic of experiments in ecology: Is pseudo-replication a pseudoissue? *Oikos* 94:27–38.

Petratis, P. S., A. E. Dunham, and P. H. Niewiarowski. 1996. Inferring multiple causality: The limitations of path analysis. *Functional Ecology* 10:421–431.

Platt, J. R. 1964. Strong inference. *Science* 146:347–353.

Popper, K. R. 1959. *The Logic of Scientific Discovery.* Basic Books, New York.

Potvin, C. 1993. ANOVA: Experiments in controlled environments. Pages 46–68 in S. M. Scheiner and J. Gurevitch (eds.), *Design and Analysis of Ecological Experiments.* Chapman and Hall, New York.

Quinn, J. F., and A. E. Dunham. 1983. On hypothesis testing in ecology and evolution. *American Naturalist* 122:602–617.

Reznick, D. N., and J. A. Endler. 1982. The impact of predation on life history evolution in Trinidadian guppies (*Poecilia reticulata*). *Evolution* 36:160–177.

Reznick, D. N., H. Bryga, and J. A. Endler. 1990. Experimentally induced life-history evolution in a natural population. *Nature* 346:357–359.

Ricklefs, R. E., and D. Schluter. 1993. Species diversity: Regional and historical influences. Pages 350–363 in R. E. Ricklefs and D. Schluter (eds.), *Species Diversity in Ecological Communities*. University of Chicago Press, Chicago, IL.

Roush, W. 1995. When rigor meets reality. *Science* 269:313–315.

Schmitz, O. J., and T. D. Nudds. 1994. Parasite-mediated competition in deer and moose: How strong is the effect of meningeal worm on moose? *Ecological Applications* 4:91–103.

Schneider, D. C., R. Walters, S. Thrush, and P. Dayton. 1997. Scale-up of ecological experiments: Density variation in the mobile bivalve *Macomona liliana*. *Journal of Experimental Marine Biology and Ecology* 216:129–152.

Shipley, B. 2000. Cause and Correlation in Biology. *A User's Guide to Path Analysis, Structural Equations and Causal Inference*. Cambridge University Press, Cambridge, UK.

Shurin, J. B., E. T. Borer, E. W. Seabloom, K. Anderson, C. A. Blanchette, B. Broitman, S. D. Cooper, and B. S. Halpern. 2002. A cross-ecosystem comparison of the strength of trophic cascades. *Ecology Letters* 5:785–791.

Smith, F. A., J. H. Brown, and T. J. Valone. 1997. Path analysis: A critical evaluation using long-term experimental data. *American Naturalist* 149:29–42.

Southwood, T.R.E., and P. A. Henderson. 2000. *Ecological Methods*. Blackwell Science, Oxford, UK.

Thomas, D. C., and D. R. Gray. 2002. Update COSEWIC status report on the woodland caribou *Rangifer tarandus caribou* in Canada, in COSEWIC assessment and update status report on the woodland caribou *Rangifer tarandus caribou* in Canada. Committee on the Status of Endangered Wildlife in Canada. Ottawa. 98 pp.

Thompson, J. N. 1999. Specific hypotheses on the geographic mosaic of coevolution. *American Naturalist* 153, supplement: S1-S14.

van Kammen, D. P. 1987. Columbus, grantsmanship, and clinical research. *Biological Psychology* 22:1301–1303.

White, T.C.R. 1969. An index to measure weather-induced stress of trees associated with outbreaks of psyllids in Australia. *Ecology* 50:905–909.

White, T.C.R. 1984. The abundance of invertebrate herbivores in relation to the availability of nitrogen in stressed food plants. *Oecologia* 63:90–105.

Wilson, D. E., F. R. Cole, J. D. Nichols, R. Rudran, and M. S. Foster. 1996. *Measuring and Monitoring Biological Diversity: Standard Methods for Mammals.* Smithsonian Institution Press, Washington, DC.

Yang, L. H. 2004. Periodical cicadas as resource pulses in North American forests. *Science* 306:1565–1567.

Yoccuz, N. G. 1991. Use, overuse, and misuse of significance tests in evolutionary biology and ecology. *Bulletin of the Ecological Society of America* 72:106–111.

Young, T. P., B. Okello, D. Kinyua, and T. M. Palmer. 1998. KLEE: A long-term, large-scale herbivore exclusion experiment in Laikipia, Kenya. *African Journal of Range and Forage Science* 14:94–102.

Zschokke, S., and E. Ludin. 2001. Measurement accuracy: How much is necessary? *Bulletin of the Ecological Society of America* 82:237–243.

Index